The Design of
Inquiring Systems

The Design of
INQUIRING SYSTEMS:

Basic Concepts of Systems and Organization

BY

C. WEST CHURCHMAN

Basic Books, Inc., Publishers

NEW YORK   LONDON

© 1971 by Basic Books, Inc.
Library of Congress Catalog Card Number 72–174810
SBN 465–01608–1
Manufactured in the United States of America
76   10 9 8 7 6 5 4 3

for
ALICE M. SANDERS
With affection and gratitude

PREFACE

A preface should explain and thank; it has two audiences, those who might be interested in reading the book and those who have made the reading possible.

The three key words of the title help to define the first set, which includes people who are interested in the philosophical issues of design, of inquiry, and of social systems. "Design" is used throughout in its most generic sense, to include planning, operations research, engineering design, architectural design, programming, budgeting, and all the other activities in which we consciously attempt to change ourselves and our environment to improve the quality of our lives. So the book could be read as a philosophy of organization theory, or of architectural or engineering design, or of operations research, or of planning.

So immodest a call for an audience needs to be bridled by a note of humility. I think we are creating a genre of books which try to look at human society in terms of manifolds of interconnected problems, and not in terms of specific problems like pollution, poverty, and power. Often these books use the word "society" to reflect this interest in the whole rather than the parts; Donald N. Michael's *The Unprepared Society* and Warren Bennis' and Philip E. Slater's *The Temporary Society* are two excellent examples. Others, myself included, like the word "systems"; Geoffrey Vickers' *Value Systems and Social Process* and Harold Sackman's *Computers, Systems Science, and Evolving Society: The Challenge of Man-Machine Digital Systems* are other examples.

Despite the fact that we are struggling to deal with the multiplicity of social problems, each of us must write with his own style, within his intellectual cell, and with his own biases. My bias has been to employ a mode of conversation with historical figures of the past who have written in a similar vein. To carry on the conversation, I've had to use a common method of communication by restating in my own language what I think the fellow was driving at; this method may very well outrage the historical scholar, to whom I can only offer apologies. Even out of misunderstandings a great deal can be learned.

The word "inquiry" suggests that the audience includes persons interested in the philosophy of science; this is true, so long as the interest is a very broad one, concerned with the meaning of science with respect to other social institutions, health, education, morality, and so on. Inquiry is by no means restricted to the disciplines of science. I could have called the book *The Design of Systems,* but this would not adequately reflect my style, which is to proceed from the more specific problem of inquiry to the "whole system."

The first part of the book develops a classification of systems, based on a conversation with five historical figures—Leibniz, Locke, Kant, Hegel, and Singer, along with their allies and enemies. The reader will naturally gain the impression that this classification is also an evolution, from the primitive forms of inquiring systems to the more advanced. Whether this impression is correct, or what it would mean to say that it is correct, is the subject matter of the speculations of the second part. Here again my personal style is to conclude with a question; the formulation of social problems is to me the most important intellectual activity.

This book has taken a long time getting itself written, so that my gratitudes are spread far and wide. The direct impetus for the first draft came from some work of Edward Feigenbaum and Julian Feldman on getting computers to perceive and think. But the more indirect impetus is a lifelong interest in the work of E. A. Singer, Jr., and his student who was my teacher, Henry Bradford Smith. Their influence has cemented a lifelong collaboration with Thomas A. Cowan and Russell L. Ackoff. Shortly after the first draft was written, Cowan and I formed a "philosophy of science" seminar at Berkeley, which had Frederick Betz and Ian Mitroff as its first "students"; the many student-teachers who have owned this seminar have had a great deal to do with the contents of this book, and especially the second part. Another interdisciplinary seminar of the Berkeley Space Sciences Laboratory was also an enormously helpful influence. All the participants are to be blamed for the inadequacies of the text. Alfred Schainblatt practically coauthored the chapter on Singerian Inquiring Systems. Other gratitudes are mentioned in several of the other chapters.

The research for this book was supported in part by the National

Aeronautics and Space Administration and the National Science Foundation.

No book by a man is written without an enormous feminine influence, which is a closely guarded secret and which everyman reveals at the end of his preface. The circumambulatory theme of Part II is the invention of my wife, Gloria; she told me the myth that the verb "to meander" comes from a Greek King Meander whose palace was a labyrinth, a myth that must be real whatever the *Oxford* says to the contrary. Robin Zoesch and Phyllis Dexter, both more astute inquirers than I in many ways, graciously typed and corrected the manuscript even though they often disagreed.

Mill Valley, California
August 1971

CONTENTS

Part

: I :

A Classification of Systems

: 1 :

DESIGN AND INQUIRY

On the Limits of the Design of Systems

Of the many activities of man, designing is one of the most fascinating and crucial. We believe we can change our environment in ways that will better serve our purposes. Some people even believe they can change themselves in this regard. What makes the activity of design so fascinating is that design enables us to create systems which will perform tasks better than a single person does alone.

Consider, for example, the activity of transportation. A strong man alone can carry, say, one hundred pounds at a velocity of three miles per hour for several hours over a smooth and flat terrain. Then someone discovers that the animals in his environment can be trained to do the same task, at an even faster rate, over a longer period and rougher country. Then insight and intelligence combine to create the wheel, and there is a system consisting of the man, the ox, and the cart, a beautifully effective system for performing the same task. By slow stages we learned to create systems of transportation that ignore the eccentricities of the ground, can adapt themselves to enormous loads, can travel several hundred or thousand times as fast as the lone man. Other examples come easily to mind: the system of shelters protects us far better than our own skin does, the system of medicines prevents disease far better than a single person can on his own.

It is natural to ask ourselves in what areas a man cannot be bettered by his own designs. What are the things a man alone can do better than any system men will ever design? Perhaps the more clever ones among us will have a ready answer: a man alone can be lonely; a system never

3

can. The answer is both facetious and serious. The serious intent is to argue that creativity is essentially a lonely process, the act of a man by himself. A man may be encouraged to perform better in certain designed environments, but he can never be designed to create. Thus, says this answer, creative arts, creative sciences, creative religions, creative politics cannot be designed; no system can ever be designed that will produce a better art, science, religion, or politics than that created by some men alone.

This answer is appealing on a number of counts. It is appealing because we are all afraid of losing our dignity as men; we are afraid that a system will "take over" and do all things better than a lonely man can do. It is appealing because we want to remain free; if a system is designed to perform better than we can, it can legitimately tell us what to do.

The answer is also appealing because it is almost, if not altogether, a tautology. What do we mean by a "creative" act? If we say an action is creative only if it cannot be analyzed or understood, then clearly no system can be designed to create, for design always implies both analysis and understanding. We shall want to avoid the pitfall of terminating all discussion by transforming the question of creativity into so trivial a response. Hence we leave it open whether creativity can be analyzed and understood. Specifically, we seek to pursue the elusive by asking ourselves where system design fails to be creative, and whether this failure is permanent or temporary.

In this essay, our interest lies in the creativity of science, i.e., in the actions that lead to new knowledge. We are interested in the extent to which man can design an inquiring system.

In order to make the purpose of the essay clearer at the outset, we need to do two things: first, explain what is meant by designing systems, and second, what is meant by inquiry. However, "explaining" either concept is, in a sense, the purpose of the entire essay. The situation is characteristic of philosophical inquiry: one wishes to discuss a concept, and hence must try to make clear what concept is being discussed, but the purpose of the discussion is to enlighten the meaning of the concept. One must necessarily put the cart before the horse. The only recourse is to begin with a tentative estimate of the meaning and allow discussion to modify the definition. Indeed, as we shall see later on, this very procedure of "endless approximation" is itself a design of inquiry.

On Design (Preliminary Statement)

The introductory remarks of this chapter themselves suggest some of the salient characteristics of design. First of all, design belongs to the category of behavior called teleological, i.e., "goal seeking" behavior. More specifically, design is thinking behavior which conceptually selects among a set of alternatives in order to figure out which alternative leads to the desired goal or set of goals. In this regard, design is synonymous with planning, optimizing, and similar terms that connote the use of thought as a precursor to action directed at the attainment of goals.

Each alternative, ideally, describes a complete set of behavior patterns, so that someone equipped with the same thought processes as the designer will be able to convert the design into a specific set of actions. Consequently, as a first approximation, design has the following characteristics:

1. It attempts to distinguish in thought between different sets of behavior patterns.
2. It tries to estimate in thought how well each alternative set of behavior patterns will serve a specified set of goals.
3. Its aim is to communicate its thoughts to other minds in such a manner that they can convert the thoughts into corresponding actions which in fact serve the goals in the same manner as the design said they would.

It will be noted that these specifications contain the phrases "attempts to," "tries to," and "aim"; the point is that the designer tries to do these three things, but may not succeed. If the phrases had been omitted, then we should have been caught in the awkward position of saying that design behavior occurs only when it is completely successful, i.e., never. Indeed, it is important at the outset to recognize that there are degrees of design, depending on a person's interest in the three efforts, as well as the amount of success he attains in them.

There is a fourth characteristic of design behavior that is important for the subsequent discussion. This is the goal of generality, or, as many would put it, methodology; the designer strives to avoid the necessity of repeating the thought process when faced with a similar goal-attainment problem by delineating the steps in the process of producing a design.

In a sense, this design goal consists in communicating with another designing mind faced with similar problems. Once the designer has had some success in this fourth effort, he can say that he can tell *why* a design is good, in addition to telling the *what, when,* and *how,* which the first three efforts attempt to accomplish. The broader the class of problems that a design methodology can be used to solve, the deeper the "explanation" of the design.

It is evident that some "other mind" is critical for the designer, whether it be his own mind later on or some different person. This other mind transforms thoughts into action (3) or into other designs (4). In understanding the design process it would be very convenient to have a standard "other mind," which the student of design could use to test the effectiveness of various design processes. A recently developed mind, the digital computer, is a likely candidate. We can defer for later argument the question whether a computer "has" a mind. For the present, it appears to be a good candidate for a standard because (a) its processes belong under the category of thinking, and (b) in principle one can test whether a set of ideas have been adequately transmitted to it.

So our question is whether it is possible to tell a computer how to design an inquiring system, or, in other terms, teach a computer to conduct research. The purpose is not to design an "automated" researcher, but rather to discover what in the research process is truly the "lonely" part, the part that cannot be designed, at least relative to a standard computer.

A "standard (digital) computer" means a machine capable of receiving discrete inputs, say, from a typewriter, and performing symbol manipulation on these inputs in accordance with specific rules (instructions) which can also be given to it via an input device. The computer can then pronounce its results, e.g., through a "print-out." The ideal standard computer has perfect replication: given the same inputs, the results will always be the same. This means that the standard computer can communicate with itself at a later time, i.e., it has a "memory." A set of inputs and instructions which operate on the inputs is called a "program."

Actually, one need not have used a computer as a standard, because a conceptual idealized logic machine would have done as well; one then has recourse to the rich literature of symbolic logic in order to precisely define "formal thinking" in terms of symbol manipulation. But the computer is better known, and there are many prototype programs which help illustrate the discussion of design.

On the Design of Systems

We are specifically interested in the design of systems, i.e., of structures that have organized components. As we move into the discussion in greater depth, we shall have to say a great deal about the concept of a system, but one central problem of all systems design can easily be illustrated. For example, the designer of a home for a family is designing a system. Narrowly, he may think of a particular instance of a design as the specification of a physical house, designated by a complete set of architectural drawings and specifications. In this case, the components may be the rooms, and the relations between the components may be the geometrical scheme of the house in three-dimensional space. But the architect may ask himself a broader question: whether the house is not a component of a larger system, consisting of the family (or its activities) and the house. When he does ask himself this question, he may wonder whether his design task should include the design of a part of the family's activities. For example, he may wonder whether he can change the family's typical way of using the kitchen facilities. Still more broadly, he may ask whether the house plus family is not a component of an urban social system, and whether he ought not to consider alternative designs of this entire community. If he perceives his task in the narrowest sense, then he tells himself that the larger system is not his concern; how the family behaves is entirely up to them, or how the community is planned is entirely up to the planners and politicians. In such a case, he believes that the maximum size of the system is the house (plus, say, its location on the land). He may believe that there is a larger system that may concern some other designer; such a larger system may be the city in which the house is to be placed. But as far as he is concerned, larger systems are not relevant to the effectiveness of his choices.

Thus, one system design problem of central importance is to decide how large the system is, i.e., its boundaries and environment. A closely related problem is one of determining the basic components, i.e., the components that do not contain subcomponents. For example, the architect may decide that there are ultimate choices he can make from a catalogue: he cannot or should not consider alternative ways of putting together the parts of a window, since this is entirely up to the window manufacturers. In this case, he regards the system to have a smallest component.

All men are system designers, and each man tries to determine what, in his world, is the largest system and the smallest. For each human, the system he designs is his life, i.e., his self. The question all of us face is what is the largest and smallest system which constitutes the self? Where does self designing begin and end?

The trouble with such a question is that it is so confusing. The intent is clear enough in each specific case, however. A man must decide whether to pay attention to his own survival and welfare only, or his family's, or his city's, or his nation's, or the world's, or of "space." He must decide whether to "take" what is offered in terms of goods and money, or to create his own. In either direction he looks for the broadest and the deepest limits of his world of system design. But to translate these familiar problems of human living into a form that can receive sensible general answers is the difficult task.

As we proceed in the discussion of the design of inquiring systems, we shall find that we must face the question of the largest and smallest system: what is the largest set of components the designer of inquiring systems must consider, and what are the fundamental components that cannot be further analyzed into systems? To illustrate, is it enough to consider just the acts of formulating hypotheses and testing hypotheses? If this is essentially all that an inquirer can be expected to do, then such matters as generalizations from theories or the communication of results are taken to be outside the purview of the inquirer, and hence matters of concern to other systems. Likewise, if the human being is regarded as essential for inquiry but the process by which he creates new ideas is taken to be forever beyond the scope of analysis, then the human creator is regarded as a fundamental component, one lower bound of the hierarchy of components.

To the four characteristics of design given above, we must therefore add a fifth which is specific to the design of systems: the systems designer attempts to identify the whole relevant system and its components; the design alternatives are defined in terms of the design of the components and their interrelationships.

On Inquiry

Inquiry is an activity which produces knowledge. This definition by itself is not very helpful unless the reader had never thought of inquiry in so broad a manner. But the definition does serve as a springboard for further clarification. By "produces" we mean "makes a difference in and of itself." In other words, for an activity to be said to produce a result, it must really matter, and to test whether it matters one determines whether the absence of the activity would have resulted in something different. As for "knowledge," we shall want to discuss the concept in great detail, because what we mean by an inquiring system depends very much on what we mean by knowledge. Thus, in a way, the purpose of this essay is as much to define knowledge as it is to discuss the design of inquiring systems.

Knowledge can be considered as a collection of information, or as an activity, or as a potential. If we think of it as a collection of information, then the analogy of a computer's memory is helpful, for we can say that knowledge about something is like the storage of meaningful and true strings of symbols in a computer. In order to better define this storage, we would then proceed to explain what "meaningful" means, and the conditions under which one could test whether storage has occurred. For example, using logic, we might assert that a string of symbols is meaningful if it is a well-formed formula in some formal language, and it is true if it meets certain semantic tests. We would go on to say that the string is stored if it can be retrieved by a specific set of operations, e.g., by querying the mind in a certain manner.

It is rather easy to imagine how this definition of knowledge leads to a very specific problem of design, and how the computer could readily play a central role as the "standard mind" in the sense of the fourth stipulation given above. Indeed, there are many enthusiasts who look forward to a national (or world) network of scientific and technological information, where men can go with their questions about nature and receive the most up-to-date answers. Admittedly there are colossal design problems to be faced in order to make such a modern library feasible, but the conceptual idea seems clear enough. But is it?

In much of the popular literature about research and science, the authors often assume that the meaning of a "systematic collection of known facts" or "collection of information" is a clear concept to most readers. Apparently they think of a "collection" in terms of a library, and a systematic collection to be like a well-run library with an adequate indexing and cataloguing system. However, no library qualifies as an entity having a "state of knowledge" in the sense discussed above. It is true that stored in it are strings of meaningful symbols. But it has no adequate way of showing which strings are meaningful and which are true. We would have to say that the state of knowledge resides in the combined system consisting of the library and an astute and adept human user. Even then, we would find it most difficult to arrive at a satisfactory test of whether such a system really had knowledge of a certain kind.

To review briefly some of the more obvious conceptual problems of a library of science:

1. Does the library speak the same language as the user? (I.e., do the user's categories exactly correspond in semantic meaning with the library's? Even between two scientists within the same discipline, the answer is apt to be a strong negative.)

2. What if the user doesn't know what question he really needs to have answered? For example, he may ask what drug is most effective in curing a disease, but may not ask for the side effects; should the library tell him? Possibly. But now suppose the user is an engineer who wants to know the strength of some material that will be used in a vehicular tunnel, but actually the tunnel itself is an economic waste of public funds. Should the library tell him so?

3. Should the library give some estimate of the quality of the information? How? That is, how can it do this without knowing a lot about the user and his purposes?

4. A national scientific library would be very expensive if it were to be available to all citizens; how should we assess the value of this public project against other public projects? That is, what are the boundaries of the inquiring system?

Thus the commonly uttered definition of science as a systematic collection of knowledge is almost entirely useless for the purposes of designing inquiring systems, because the definition fails in all ways to provide any clue concerning what the inquiring system is supposed to accomplish. "Science" is certainly not the Library of Congress, any more than medical science is the National Library of Medicine plus the Index Medicus. But what more is science besides these "systematic collections"?

Put otherwise, to conceive of knowledge as a collection of information seems to rob the concept of all of its life. Knowledge is a vital force that makes an enormous difference in the world. Simply to say that it is a storage of sentences is to ignore all that this difference amounts to.

In other words, knowledge resides in the user and not in the collection. It is how the user reacts to a collection of information that matters. Hence we should turn to the other concepts of knowledge, action and potential action. The action conception of knowledge is pragmatic; knowledge is an ability of some person to do something correctly. The person exhibits a form of knowledge if he can perform an assigned task correctly. But this is only a very restricted notion of the class of actions representing knowledge. In the first place, we would not want to argue that an entity has knowledge only when it is acting. A carpenter knows how to frame a window even when he's sleeping. But this is a familiar enough condition of many physical objects: a copper wire is a conductor of electricity even when no current flows through it. Indeed, only rarely do the objects of the world become tested for their properties, so that almost all of them are described in terms of what they might do under certain circumstances. Thus knowledge is a potential for a certain type of action, by which we mean that the action would occur if certain tests were run.

For example, a library plus its user has knowledge if a certain type of response will be evoked under a given set of stipulations, e.g., a correct sentence given a certain type of question. This way of conceptualizing a collection of information is far more useful from the design point of view than thinking of a library alone as a collection; a library so designed that the retrieval of information is either impossible or much too time-consuming is *not* a collection of information, no matter how many correct sentences are stored there. It is not a collection, because it fails to provide the correct response, given a query.

The vitality of knowledge resides not only in its being merely a potential, but a potential of a very powerful sort. It is not enough to perform correctly—many ignorant minds do this on occasion by the laws of chance. To be knowledgeable, one must be able to adjust behavior to changing circumstances. The lesson is a familiar one. All of us on occasion will treat another human being with understanding and compassion; it takes a lifetime to learn how to treat a human being with understanding despite changes in mood, changes in economic circumstances, and the advent of tragedy. Any man can learn to "sail" in one fine afternoon; it takes a lifetime to learn how to sail. Knowledge is being at once at ease with a subject and deeply engrossed in it. Knowledge carries with it both a tremendous joy and a great despair—a joy of being at one with a whole area of living human activity, and a great despair in recognizing how little this oneness really is compared to what it might be. Nothing touches the true depth of the human spirit so much as the act of knowing.

Perhaps no greater poetry in praise of knowledge was ever written than Spinoza's *Ethics*. The point of the poem was not only to dignify once more the ability of man to know, but also to portray this state in all its magnificence. The smaller mind of the early learner is constantly worried about its freedom to do many things; it abhors the tedious discipline of learning calculation, learning scales, practicing the steps. It wants above all to be able to do something else, something other than what it is now being forced to do. The larger, knowing mind at last attains the state of desiring to do the same thing no matter how the situation changes. But it also comes to realize that "doing the same thing" is not a rigid routine, mechanically performed. Instead it is the ability to pursue what one most deeply desires, to express what one fully realizes needs to be expressed, no matter how difficult the circumstance, how tedious the task, how tragic the mood, or how joyful the occasion.

Despite the wonders of modern science as a creator of knowledge, men have found it very difficult to determine how they should feel about knowledge. In Spinoza's day, knowledge, i.e., pure truth, was a marvelous accomplishment, to be glorified in its own right. Today we are highly suspicious of what science produces, and are not in the least comforted by the spokesmen for science who still proclaim its purity of heart. Indeed, we have made up a term, "scientism," which would sound peculiar indeed to Spinoza's age: it means the attempt to reduce all matters of concern to science. If "science" means intellectual understanding, then Spinoza would be puzzled and horrified to learn that modern man is afraid of scientism, i.e., afraid of "reducing" his most serious problems to intellectual understanding. Clearly our "science" does not mean what Spinoza took understanding to mean. It does not because, in the course of events, science left out of its concern certain ingredients that men of Spinoza's day would have regarded as essential for understanding: moral value and God. We have a way of saying that a weapons system is "good" or someone is a "good" manager, without meaning to imply that either has moral worth or ultimate intrinsic value. For Spinoza, without a clear concept of the ultimate human values, one cannot appreciate understanding, and without a God one cannot evaluate understanding.

To the designer of an inquiring system, these remarks are quite relevant. He must wonder whether the system is to be designed to produce knowledge of the sort that present-day man calls knowledge, or of the sort that Spinoza called knowledge, or something else. If knowledge means the ability to pursue goals though the world about us changes, then perhaps an inquiring system that produces "science" does not produce knowledge. There seems to be sufficient evidence to make the designer at least pause long enough to consider this issue. There is no way to consider it except to permit some breaking away from present practice; perhaps, as I have been hinting in resurrecting Spinoza, we need to turn toward a reactionary policy. In any event, the designer must let his feelings, as well as his common sense and thought processes, tell him some things.

On Design and Implementation

It can be seen that design, properly viewed, is an enormous liberation of the intellectual spirit, for it challenges this spirit to an unbounded speculation about possibilities. The student of modern science is constrained to the boundaries of accepted practice: what scientists do, try to do, expect to do, can do. He has been trained—rather than educated —to assume that "inquiry" must be limited within these constraints. He also takes the method of inquiry to be a reconstruction of scientific practices, a reconstruction that eliminates inconsistencies arising out of the confusions of individual scientists, and he tries to build a logically satisfactory theory of verification. But this reconstruction is based on practice, not on imagination.

The liberated designer of inquiring systems, on the other hand, will look at present practice as a point of departure at best. A liberated designer of vehicles can wonder whether men should transport themselves and their goods in the awkward way they now do. Such a designer asks himself why some people need to dash around over the earth's surface. He wonders, "What is the largest transport system? Does it include communications as well? If so, can we design a system where a person has a choice as to whether he travels or telephones? Could we more economically hold television conferences? But then why do men have to confer? Does the transport system include social and psychological aspects of human behavior? If so, what is the optimal design of a conference?" And so on.

The intellectual freedom that belongs to the designer must be paid for at a considerable price. We shall want to ask ourselves what is the largest inquiring system, and this question will certainly take us far beyond the limits of modern science, its laboratories, and its libraries. But in the end, we, like all designers, must come back to face reality. This means that we must ask: What can be done about it?

The transport designer may decide that yes, the communications system is included in the transport system. Yes, people travel too much, and television could easily eliminate the need for so many trips. Yes, let's get the federal government to subsidize large-scale research into feasible televised conferences. Everyone knows the consequences of such a suggestion: the airline, bus, and rail companies won't see it that way at all. The consequences will be political as well as intellectual.

The situation reminds us of the many suggestions made by academic minds as to how to solve the world arms question. If Nation A would destroy one weapon unilaterally, then Nation B would follow suit; if Nation A would put up a bond with the United Nations or designate a group of its most prominent citizens as hostages, then no aggressive act could occur. What the suggesters failed to realize—or did not know how to realize—is that the human being does not move from one kind of activity into another just because on some grounds the latter appears more "reasonable" than the former.

These remarks suggest that the definition of design must be enlarged. Design, according to the first two stipulations given above, is primarily a thought process. The third and fourth stipulations add communication, and say that a successful design is one that enables someone to transfer thought into action or into another design. But suppose that "someone" doesn't want to make the transfer of the design into action, even when thought says that the action will serve his own best interests? Suppose the "someone" is against change even when thought says he ought to change?

Should the designer of systems include resistance to change in his design? To fail to do so seems to leave out the most important aspect of the problem. But how to include resistance to change is not at all easy to see. Often the redesigner of a system—like an industrial firm— would be wise if instead of first asking what's wrong, or what changes would create the greatest cost savings, he asked himself what can be changed, and how? It takes only one or two days of work in many federal agencies to learn about practices that are unnecessarily time consuming; it takes several months of frustration to learn that this information about the system is useless because the obvious changes can't or won't occur.

What hope for change is to be expected by the designer of inquiring systems? At first blush, very little. Scientists tend to develop very strong opinions about the validity of their own methods, as every critic and adviser soon discovers. Like most decision makers who hold some position of prestige and rank, they classify possible change into two kinds, the sort that will threaten their position and the sort that will not. I can tell a researcher about a new article that he might not have read, and he will not only be grateful, but will go and read the article. I can also tell him about some work done in another field that he would not normally be expected to know. But if I tell him that his basic method is bad, or his facts unreliable, or he ignores relevant sources in his own field, I tell him that he doesn't deserve the position he has.

Quite naturally, scientists tend to resist changes in their basic outlook on the world. They find reinforcement from their fellow workers. It would be futile indeed to suggest to our scientists of today that they've probably been concerned with the least important and easiest problems of nature, no matter how valid the statement might appear to the designer. Only on matters that do not threaten their role do scientists show a ready inclination to adopt change.

Thus it seems to accomplish very little for the designer of inquiring systems to say that scientists ought to spend much more time on the study of the brain, or of disease, or of social systems, or of human wants and needs. The scientific community, far from accepting them, is simply not organized even to consider such suggestions in a systematic way.

But if the designer suggests better ways of indexing and abstracting, or in general suggests some ways of improving channels of communication in science, then both the organizations required to consider the proposals and the necessary deliberative attitude are available. Few scientists will feel any threat to their roles; now the resistance, if any, will come from the current custodians of documents: editors, publishers, and librarians.

It almost seems as though science can change only in trivial ways, or only in rather dull technical, if important, ways. The designer can color the walls a different shade or rearrange some of the passageways between the rooms, but the basic structure remains.

There is one consideration, however, that may lend a shade of optimism to this gloomy picture of the designer's task. Problems of allocation of resources always suggest to the designer that he may be able to eat his cake and have it, too. This seems especially to be the case in the allocation of time, when the designer goes to work to determine whether the same task cannot be accomplished in less time, or better accomplished in the same time.

In other words, the problem of inquiring systems may not be so much the allocation of personnel and budgets to various areas, but rather the improvement of the process of research itself. Consider the possibility that 50 percent of a skilled researcher's energy is spent on activities that an intelligent technician could accomplish just as easily, with perhaps more accuracy. Can we design research organizations so that a more satisfactory allocation of this precious energy is accomplished?

This is a very subtle question, and perhaps a badly posed one, since it appears to make the questionable assumption that researchers carry around so many units of energy which they can allocate to various tasks. The more general issue is whether research activity can be vastly strengthened by strengthening its support activities. We are all familiar with various forms of these activities: literature search, comparison of concepts and ideas, mathematical modeling, computation, reporting, etc.

On Intelligent Technicians

The term "intelligent technician," introduced by Allen Newell, seems a very apt phrase to describe this sort of support activity of research. It may be a good name for the "standard computer" discussed above. If we turn our attention to the design of such technicians, we may find a way to consider the whole design of inquiry that permits both wide speculation and realism. Our interest will not only be confined to the intelligent technician because we shall be trying to understand the nature of the broader system that includes the technician as a component.

The realism of the design of the technician's role lies in the possibilities of computer design. We shall be asking ourselves to what extent we can instruct an idealized computer to perform support activities for research. But this is not a book on detailed design as such, but a prolegomena to all such designs. Our interest is in determining how to regard such technicians. Specifically, what tasks of research can they perform and what tasks are forever beyond their capabilities?

We can easily estimate some answers on the positive side: these technicians already do a reasonably good job of storing and retrieving information, condensing information, building simple and sometimes complicated models, computing at fantastically high speeds, printing out results, and so on. At present the computer technician is a clumsy animal, hard to consult readily, and expensive, but all these defects are temporary. The interesting question for speculation on the designer's part is what this technician cannot do.

As was pointed out above, the obvious reply is unfortunately a tautology. It is the reply that the intelligent technician cannot create new ideas, cannot make innovations. Possibly, though not certainly, the technician could not have written this book. The reply is a tautology because what is meant by the creative act is an act that cannot be designed beforehand, although it may be analyzable in retrospect. If this is the correct meaning of creativity, then no intelligent technician can be creative. But this conclusion is absolutely worthless for the designer of inquiring systems. What he needs to know is the method of identifying the creative act. It does not help the designer when the creator says that he doesn't know how he came upon a given idea, nor is the creator's inability to describe his process of creation any evidence that his innovations cannot be designed; the creator may not have the ability to reconstruct his own thought processes.

Thus the attempt to have our whiskey and drink it too can perhaps be accomplished if we keep an eye on the design of an intelligent technician who will minimally threaten the establishment, while at the same time we let speculation take us far beyond what scientists do or are apt to want to do in the foreseeable future.

Historical Designs of Inquiring Systems

How shall we proceed? As I suggested above, a reactionary process of discussion offers some very attractive possibilities. The current tendency in designing inquiring systems is to bolster science and its research as it is conceived today. But in every age when men have struggled to learn more about themselves and the universe they inhabit, there have always been a few reflective thinkers who have tried to learn how men learn, and by what right they can claim that what they profess to learn is truly knowledge. This is reflective thinking in the literal sense: it is the thinking about thinking, doubting about doubting, learning about learning, and (hopefully) knowing about knowing. If we accept the thesis that these reflective minds did indeed learn about learning, then their contribution to knowledge is quite important for the designer of inquiring systems.

In a way, we can regard the history of epistemology (theory of knowledge) not as a description of how men learn and justify their learning, but as a description of how learning can be designed and how the design can be justified. This way of reading the older texts requires a translation, not from one language to another, but from one philosophical aim to another. We are less interested in what Leibniz, say, was trying to accomplish than in what his attempts mean to the designer.

Therefore when we speak of a Leibnizian inquiring system, we do not mean that this system is an exact account of how Leibniz conceived the theory of knowledge; rather, it is a reconstruction of Leibnizian ideas in the language of the design of an inquiring system.

In the case of each writer, we shall follow enough of what he has to say to be able to construct a design of an inquirer, and we shall see what he or others have had to say about how satisfactory or unsatisfactory such a design really is. Since we are interested not only in the design of an intelligent technician, but also the broader questions of the design of creativity and the limits of the inquiring system, our investigation will take us far beyond what we can feasibly put on a computer at the present time.

It will not hurt to say where this historical journey will lead. It will suggest some very rich and exciting designs, each design in some sense encompassing the best features of its predecessors. But in the end it will conclude that we are faced today with some critical design problems we do not know how to solve. There will be the suggestion that science's mode of representing nature is very restricted, so that it cannot even talk about some of its most pressing problems and specifically its relationship to other social systems. For example, science has no adequate way of studying its own relationship to politics, to religion, or even to a system apparently quite close to its own interests, education. As a system, science cannot discuss social change (implementation) in any but a very restricted sense. Finally, and perhaps most important, science has no adequate way of studying the elusive, since it always aims for precision, and hence in some real sense science is alienated from nature.

The particular historical figures chosen in this exploration are in part a personal choice. Plato, Aristotle, or St. Thomas would certainly be reasonable candidates. But the renaissance of epistemology in the seventeenth century has seemed to me to make a better starting point because everything was so open to speculation and imagination. One of the greatest men of this age, René Descartes, could honestly call for a clearing of the slate and the design of an inquiring system *de novo,* which, after all, is what this chapter has asked for.

: 2 :

LEIBNIZIAN INQUIRING SYSTEMS: FACT NETS

The Inputs of Inquiring Systems

Many persons who have thought about the methods by which men gain knowledge have come to the conclusion that the mind begins by learning the simplest and clearest things first, and then proceeds to learn about complex matters by "building up" from these elementary forms of knowledge. Thus the student learns geometry or astronomy by first being introduced to the "elements" and then proceeding to more advanced topics. Von Neuman and Morgenstern, when discussing the implications of the theory of games, suggested that a two-person, zero sum game is analogous to the two-body problem of astronomy, and that with diligence we may move from the elementary competitive phenomena to a precise formulation of "real life" economic competition. However, from the point of view of the design of an inquiring system, it is by no means obvious that the elementary-to-complex process is optimal, no matter how attractive it may seem at first glance.

The second feature of inquiring systems that many thinkers take to be obvious is that the system's learning begins with an "input" or, as older writers would say, with sensory experience of the outer world. Thus the so-called realist abhors the abstractions of introverted minds which weave the webs of imagination and fantasy, no matter how refined, precise, and regular the fabric may be. But here again the designer of inquiring systems must not let himself be led unconsciously by the forcefulness of common sense. The design question is whether the ultimate origin of the material the inquiring system processes must be considered as an input, i.e., as a "given" which the inquiring system cannot decide about or control. If inputs are defined in this way, it

19

is by no means obvious that "inputs" are optimal in the design, because the inquiring system may lose control at the very place where control is most needed, namely, the origin of the "matter" (sense data, idea, or whatever) which it processes.

If we join the two concepts of the design of an inquiring system, we see that there are four pathways the designer may follow. He may so design the system that: (1) it begins with elementary inputs which are clear and distinct; (2) it begins with clear and distinct ideas which are not inputs (i.e., not "given" externally); (3) it begins with unclear inputs; or (4) it begins with unclear material which is not an input.

In this and the next few chapters we shall see that there are very convincing arguments to show that the first three alternatives are unsatisfactory from a design point of view. In this chapter we shall reexamine the epistemology of seventeenth-century rationalism in order to see why choice 2 is inadequate for those designers who wish to avoid inputs, and in the chapter on Lockean inquiring systems we shall see why choice 1 runs into design difficulties. Choice 3 will be discussed in detail in the chapter on Kantian inquiring systems; its difficulties set the stage for much richer designs along the lines of choice 4, which will occupy the rest of the book, beginning with Hegelian inquiring systems.

It will be seen then that our historical excursions have as their main purpose the establishment of a design base, namely, an inquiring system which is open as to its beginnings and which has control over all the material it processes. Two points should be kept in mind. First, it should be reemphasized that "input" has a very specific meaning here, in terms of where the control of the origin of the system's material lies, within the system or outside. Second, since we shall be arriving in this chapter at a system (called "Leibnizian") which accepts choice 4, this system will be the base on which all subsequent systems (beyond the Kantian) will build: all subsequent designs are "Leibnizian" in one very real sense.

In this discussion I have talked about the "material" which the inquiring system processes. Hence I have implicitly assumed, and now for the next several chapters will explicitly assume, that the inquiring system is a processer, and specifically a symbol processer. The symbols may have many different forms: they may be sentences, or "codes," or a set of digits, or "pictures," or some other images. In addition to identifying the symbols, the processing of the inquiring system includes storage (in memory) and retrieval (recall from memory), and the combination, transformation, and breaking up of symbol clusters. The first design concept introduced at the outset of this chapter says that the inquiring system should be so designed that it can examine each symbol or cluster of symbols and determine whether it is (a) simple or complex, and (b) clear or not clear. To be specific, if the symbol is a sentence, and the sentence is simple and clear, then the thesis asserts that the inquiring system can determine directly, without reference to anything else, whether the sentence is true or false. The design choice is to have the inquiring system begin by selecting those sentences that are simple and clearly true, which it subsequently combines to develop more complicated truths.

In order to develop this design thesis, we need to answer two fundamental design questions: (1) where do the symbols (sentences) come from? and (2) how does the inquiring system know that its direct classification of a sentence into "true" or "false" is warranted? In this chapter we shall be concerned mainly with the second question—the warranty of the simple and clear truths—although inevitably the discussion will lead us into some ideas about how to answer the first question.

Now the second question presupposes that the inquiring system can identify simple and clear items, and the designer needs to know how to design the system to do this. It is true that the human mind often seems quite capable of performing this task. If I offer the following four sentences, many human inquiring systems would have no trouble at all in classifying them into the simple and the clear: "2 + 2 = 4"; "This patch is blue"; "The quality of mercy is not strained"; "The acceleration of a falling body *in vacuo* is constant." The first two sentences are candidates for being simple and the first is a candidate for being clear and true. Assuming that the human mind can perform this kind of exercise, how does it do it? It should be emphasized that we cannot say that human minds "obviously" have the capacity of recognizing simple and clear inputs, because this statement provides no clue for the designer, i.e., the person who wants to know how the mind has this capacity.

Even if the problem of simplicity can be solved, there is still the additional problem of determining which simple and clear items are true. In other words, what guarantees that the processer that stamps "truth" (or "falsity") on simple and clear items is working correctly?

To study the questions of simplicity and truth, we turn to historical resources of ideas, and begin, as men so often do on these matters, with Descartes.

The Cartesian Guarantor

Suppose we translate a portion of Descartes' Meditation III of his *Meditations on the First Philosophy* not from French to English but from a descriptive account of a man's reflections to a theory of the design of inquiring systems:

"Why is it not sufficient to say that the designer of an inquiring system need only design into the system the things that are very simple and easy to understand in the sphere of arithmetic or geometry, e.g., that $2 + 3 = 5$? But the designer has some reason to question these 'givens,' because for all he knows he has designed a system in which the inputs originate from an unreliable source. For if the designer simply permits the system to accept the so-called clear truths of arithmetic and geometry as inputs, then it is easy enough for him to conceive that the input device causes error even though the system takes the error and classifies it as clearly correct. . . . In order to remove this fundamental defect of the inquiring system, the designer must design the system so that it can guarantee the validity of what it takes to be clearly correct sentences. Thus the designer must design the system so that it can prove that (1) there is an origin of the sentences and (2) this origin cannot produce sentences that are taken by the inquiring system to be clear and true and yet are actually false. Inability to solve these two system-design problems means a failure to design an inquiring system at all." It will be noted that Descartes does acknowledge the need for an origin of the sentences of the system, but since the inquirer can control the sentences it accepts (in terms of their clarity and validity), the sentences are not "inputs" in the sense given above.

The point that Descartes makes is both fundamental and relevant to any research into the design of inquiry, including problem-solving machines: is the system capable of guaranteeing the validity of its own results? Consider, for example, the Samuel checker-playing machine (1963) that seems to perform reasonably well in the task of playing checkers. Does the machine know that the rules by which it plays are valid? The answer is rather subtle, and in the long run may be "yes," because the Samuel checker player need not rigidly hold onto a strategy; to some extent it controls the rules by which it plays. On the other hand, most existing chess-playing and problem-solving machines do not control the rules (Newell, 1963, 1963a; Gelernter, 1963). Finally, the Samuel checker player begins with the rules of checkers, which it does not control.

At first sight it seems to be an outrageous requirement that the designer of an inquiring system design the guaranteeing component, because this seems so much the task of the human mind. But this book is not chiefly concerned with machines as such. That is, the designer is not constrained to the use of machines as his only resource. If no machine has been built comparable to the human brain, or even the monkey brain, then the designer can surely design his system with human or ape minds if he so chooses. The Cartesian question still remains: how to design the man-machine system so that the whole system can guarantee its direct method of certifying the truth and falsity of simple and clear proposals.

God as the Guarantor

Unfortunately, Descartes does not provide a very helpful answer, though his argument is surely worth repeating. The inquiring system, he says, must be designed with a capability of showing: (1) that the ultimate origin of its symbols is God, if He exists; (2) that God exists; and (3) that God is never a deceiver. The first assertion in the hands of Descartes is a tautology: God is defined as the ultimate cause of all things. Thus the inquiring system can prove this assertion if it has the capacity to define God in this manner and if it knows that tautologies are true.

The second assertion is a much more difficult matter to analyze historically, as we shall see. The problem is to determine what the inquiring system needs to know in order for it to know that God exists. Briefly, Descartes seems to argue that it must know that every event has a cause, and that the causal chain is finite ("bounded" in mathematical terms). The upper bound of the chain is defined as "God." Assuming that the inquiring system knows that "causes" is transitive and asymmetric, then it knows that God ultimately causes all items of its symbol stream, and God's actions are never caused by anything.

This does not yet imply that God is unique. Hence the inquiring system also needs to know that the upper bound of two causal chains must be the same thing.

The third requirement, to prove that God is not a deceiver, is even more difficult, for there seems to be nothing inherent in the definition of a first cause that would guarantee its reliability. Thus the inquiring system must somehow know that the unique upper bound of all causal chains is benevolent. Furthermore, there needs to be a knowledge that the causal chains ending in simple and clear inputs to the inquiring system are not distorted, i.e., that what appears true is in fact true. Thus the Cartesian inquiring system needs to have an ability to know many things that today's science would seriously question.

Furthermore, Descartes' inquiring system can somehow identify the simple and clear sentences, but how it does this is not apparent at all. We can suspect, for example, that $2 + 3 = 5$ is not simple, but a fairly complex assertion built up from a set of "simpler" sentences in arithmetic. Anyone who has worked in the fascinating area of "axiomatizing" a formal, deductive system becomes quickly aware of the enormous number of design choices in the selection of the axioms. There are at least a dozen axiom sets for Boolean algebra, for example, and it seems a matter of taste, not knowledge, as to which begins with the simplest and clearest ideas. Nicod (1917–1920), for example, reduced the set of axioms for the sentential calculus to one sentence, but no one could claim that this sentence is clear and obvious.

Finally, it is worth mentioning in passing that Descartes did try to invent a design method—the method of doubting—that would lead to some unquestionable simple truth. But later reflection of a logical character shows that the method is defective; as long as we insist that the inquiring system meet elementary requirements of logic, the Cartesian method seems inapplicable.

Nevertheless, it would be foolish to end this discussion on such a discouraging note. People do resort to the Cartesian method in religion, morality, and politics without much question. If one says that the Bible must be right because it is the word of God, then he is following the line of attack for religious inquiring systems that Descartes suggests. One begins by identifying the origin of religious statements, satisfies oneself that the origin is not deceptive, and builds a religious doctrine on the simple, clear, and valid truths. The clash between religious and so-called scientific inquiring systems is well known, nor is the debate settled except in the minds of strong believers. But from the point of view of the open-minded designer, it is as yet very difficult to see how to implement a Cartesian program, and hence the designer seeks to see whether there are not more understandable design choices.

Spinoza's Intuition

As has been mentioned, if the inquiring system could operate as a formal science like geometry, then its "beginning" must be some set of axioms. To avoid the debate about which axiom set is simplest, suppose we give up the requirement of beginning with simple sentences and address ourselves to the question whether an inquiring system can be designed that will accurately identify the true axioms. We shall say, following Spinoza, that the inquiring system has a processer that *intuitively* accepts certain assertions.

Mathematical economics seems to be a good example of such an inquiring system. Consider, for example, the axiom which asserts the transitivity of preference: if A is preferred to B, and B to C, then A is (or rationally must be) preferred to C. The intuitive appeal of this assertion is so great that few if any economists feel the urge to build formal economic systems in which the axiom fails. Geometry, of course, is the classic example; the intuitive strength of Euclid's "postulates" was so great that for two thousand years geometers played their games strictly within the domain of geometrical relations which Euclid laid down. Even when non-Euclidean geometries were discovered in the early nineteenth century, most mathematicians never thought of them as valid.

What characterizes the intuitive appeal of certain assertions? Could we design an inquiring system with intuition? Spinoza's genius suggested a clue, what one might call the "recursive" property of intuition: the intuitive faculty is such that when it directly accepts a proposition as true, it also directly accepts that it knows it to be true. In order to understand this idea better, we can recall Spinoza's taxonomy of inquiring systems. There are four types, the first three requiring inputs.

The *first* method of inquiry he calls "hearsay" (*ex auditii*). This is the kind of "knowledge" a computer has when information is programmed into it and stored in memory. There is nothing in the computer to guarantee that this information is accurate, but the computer can retrieve it on request, sometimes in fairly subtle ways. A good example is Synthex (Simmons, 1960), which has stored a small encyclopedia in a computer, and which answers elementary natural language questions. The entire encyclopedia is an input because Synthex cannot guarantee the validity of the information stored in it, and the system will respond inaccurately if the encyclopedia is wrong, without any feeling of inadequacy on its part. On the other hand, Synthex can evaluate answers to questions in terms of its own storage. For example, it can assert that one of its proposed answers is "not very relevant" or "not very accurate" in terms of what it has stored in its memory.

The *second* method Spinoza calls "vague" experience. This is "experience that is independent of the intellect," i.e., independent of cognition. A good example is the pattern-recognition machines. The "lowest" form of these machines recognizes patterns only if they are exactly like a standard within a specified morphological range, for example, contain lines that intersect at specified points on a grid. The more sophisticated forms act more like the perceiving eye, and can extrapolate, or otherwise fill in unspecified gaps, and thus recognize a badly written A, for example. EPAM (Feigenbaum, 1963) is at a still higher level in this category, since it can subclassify patterns into parts and other properties. Spinoza would not permit the term "intellect" to apply to these machines, because they do not contain any device that permits explanation. Thus even though EPAM can memorize nonsense syllables, it cannot ask why a given input occurred with a recognizable set of properties. It has no intellectual curiosity.

The *third* type of inquirer has "knowledge that arises when the essence of a thing is deduced from another thing, but not adequately." Examples are problem-solving and game-playing machines. The machines can examine a proposal, tell whether it is likely to be true or false, and attempt to prove the truth or falsity from a set of given axioms or rational principles. The problems may be of the Sunday-supplement type, or serious problems of a branch of mathematics, or the playing of a game like chess or checkers. Spinoza also seems to include in this category machines capable of inferring the causes, i.e., the explanations of events, and hence all the so-called "induction machines." An induction machine attempts to discover a sentence that will explain a whole series of events. The relation "explains" is richer than the relation "implies"; for one thing, "p explains q" is true only if p itself is true, whereas a false proposition may imply a true one. However explanation is defined, the induction machine restricts its activity to explaining *given* events in terms of alternative *given* explanations. By this I mean that the machines do not investigate the givens, nor do they attempt to chase back along the chain of explanations for an ultimate explanation. Nor do the problem solvers try to justify the rational principles with which they start, e.g., the Euclidean number system or geometry. Thus the system can "prove" that there is no largest prime number, but it does so by accepting certain "fundamental" properties of all numbers as inputs, or it can "prove" that $a^2 + b^2 = c^2$ in a right triangle by accepting Euclidean axioms as inputs.

Suppose we introduce the term "executive" for that part of an inquiring system that: (1) determines the functions of the other parts; (2) determines which part should be used in a given circumstance; and (3) judges the adequacy of a part's performance. Then in Spinoza's first class of inquirers, the executive is pretty much what programmers mean by an "executive routine" for many familiar programs. In pattern-recognition machines the executive function is more difficult to identify, but essentially it is the activity of deciding whether to modify a previous routine for classifying items. (Pattern recognizers also have the first kind of executive routine built in them.) The executive is therefore some kind of a heuristic routine. Similarly, in the third type of inquirer, the executive decides whether a proposed method of proof is likely to succeed or not. But the executive is constrained, in that he has no authority to question the inputs (what is given) or to create new methods of proof using revised rules of deduction.

Could an inquiring system be designed with a completely free executive? If so, we would have an example of Spinoza's *fourth* inquirer, whose characteristics are described as follows: "There is the knowledge that arises when a thing is perceived through its essence alone, . . . and a thing is perceived through its essence alone when from the fact that the inquirer knows something, its executive understands why it knows it." That is, the executive has a valid theory to explain why knowledge occurs.

Spinoza's classification is particularly helpful with respect to the aims of this book, because I am interested in the extent to which one can approach the design of an inquirer like Spinoza's fourth type. This seems to me to be a far more fruitful question than whether one can design thinking machines or intelligent machines. It is more fruitful because we don't even know whether our own minds are designed to accomplish this fourth type. In other words, it seems far more useful from a design point of view to ask whether one can design a system to conduct inquiry with a free executive rather than whether one can design a system to simulate the human mind.

However, although Spinoza's classification is excellent, one obtains little help from him in determining how the fourth design problem is to be solved. Spinoza does conceive of his executive as a single operating unit that uses "intuition" as its mode of functioning. A search of Spinoza's text, however, only reveals a frustrating circular language. The executive's function is "to know that the inquirer knows" (and, of course, to know that it knows that it knows, etc.). But if one asks how to design such an executive, the text seems merely to tell us that we design something that recognizes essences, i.e., so recognizes a thing that it knows why it knows. The fault is not Spinoza's, of course, for he thought that geometry was an example of a science in which his executive had been successful, and therefore that this success could be extended to other fields of knowledge.

Nevertheless, as in the case of Descartes, the design idea is still there and should not be discarded simply because intuition is so elusive a concept. All scientists learn to beware of their intuitions because they often turn out to be faulty. A great deal has been written, e.g., by Hadamard, Poincaré, and Polya, on intuition in mathematics and the important role it plays in creating new mathematical ideas. These writers all seem to agree that intuition is a kind of unconscious thinking. If so, they are not talking about Spinozistic intuition in his fourth type of inquirer, but rather the deductive process of the third type. Intuition in the fourth type must above all be conscious, because as it grasps the truth it immediately reflects on this action and simultaneously verifies that it knows. It intuitively answers the question, "Do I really know that this is so?" at the same time that it is answering the question, "Is this so?"

Furthermore, Spinoza's intuition is a very common experience in human beings. It occurs, for example, to the student who has been applying some principle his teacher has announced, without really seeing what the principle means. He has been acting like Spinoza's third type. Suddenly, he understands the principle and sees why he understands it; his "intuition" is activated. What the designer cannot yet comprehend is how to create such a faculty or recognize its existence. Even if intuition were occasionally faulty, it would be a very significant addition to an inquiring system if we could see how to formulate its properties more precisely.

After Spinoza, the idea of a free executive as a definitive truth-identifying component is discarded in rationalism. The design problems therefore become much more complicated.

The Leibnizian Inquiring System

In Leibniz, the learning process of the inquirer does not begin with clear and distinct valid truths. Suppose the inquiring system has the capacity of identifying sentences, and suppose it has an ability to apply the fundamental laws of logic. If these two components of the system can be designed properly, then the inquiring system can determine which sentences are tautologies, which are self-contradictory, and which are neither. Leibniz calls these last "contingent." As we shall see, the processing of these contingent truths is the critical problem of the design; but the point is that a contingent truth need not appear obvious or clear to the inquirer. It is the relation of the contingent truth to other contingent truths that matters.

In a very special sense, however, the tautologies and self-contradictions seem to be clear and distinct in Leibniz's system. We can begin by considering how an inquiring system would identify these. It does so by storing a set of definitions—a precise dictionary—and determining whether a given sentence follows logically from the definition. If so, the sentence is a tautology. A simple example is, "Too much eating is bad," where the dictionary defines "bad" to mean "deleterious excess," and goes on to define "deleterious" and "excess" as "too much." A more subtle example is the definition of an Euclidean straight line in such a manner that it implies Euclid's Fifth (Parallel) Postulate.

The great advantage of Leibniz's "logic processer" is that it need only analyze the form of the sentences proposed to it. Indeed, Leibniz seems to have recognized the urgent requirement for an abstract processer of this sort, and to have seen that ordinary language texts confuse the processer. He set to work trying to develop a universal language, with a structure sufficiently precise so that the processer would not be confused. For example, if we ask the logic processer in English whether all blackbirds are black, it will consult its English dictionary and find that blackbirds are wild birds with dark plumage and a very sweet note. After a bit of tortuous tracking down of "dark" and "sweet," it will place the sentence among the contingencies, albeit with a bit of a worried look on its face. No such ambiguity occurs in the universal language, where the question has to be asked in the form of the algebra of logic, e.g., "Is something that is both black and a bird, also black?" Or, more precisely, the inquiring system functions with a specific set of logical operators and relations, and expresses everything in the logical form. Thus the processer takes the sentence "black and bird is included in black" and maps it into the abstract form $a \times b \supset a$ ("what is both a and b is included in a"), which it then identifies as a tautology. Similarly, "green apples are not apples" becomes $\sim [a \times b \supset b]$ ("it is false that what is both a and b is included in b"), which is self-contradiction.

But, as we shall see, there is as yet no foolproof method of identifying tautologies so long as the inquiring system has a language as rich, say, as English. In the modern Leibnizian inquiring system, all sentences are contingent.

The Leibnizian Processer

In order to make the operations of the Leibnizian processer clearer, we can employ some logical theory not available in Leibniz's time. We imagine a system capable of handling a stream of symbols. The stream can be broken down by the system into a sequence of elements. Such an ability implies that the system can tell which elements came before which others; i.e., the system can order the elements in time. Just how it succeeds in identifying elements and ordering them is a matter for later discussion. In Leibniz's terminology, the stream is called *perception*. Perception includes the functions of generating the stream and identifying the elements. A critical aspect of Leibniz's inquirer is that perception does not originate from outside the system. In the monad, which is Leibniz's inquiring system, the perceptions are generated from within the system. This does not affect the internal operation of sorting the perception stream, but it does have some very relevant implications for the design of the whole system, as we shall see.

In higher forms of monads, the system can operate in various ways on the perception stream (the function that so operates Leibniz calls *apperception*). Specifically, the system can store segments of the stream in memory, with an appropriate tag to indicate when the segment occurred. Furthermore, the system can retrieve segments from memory and combine them in various ways (by means of *imagination*).

Perhaps the most difficult elementary task of the inquirer is its ability to identify meaningful sentences in the stream, or to construct sentences out of segments stored in memory. This could be done if the stream consisted only of symbols in a formal language, and the system had a processer that contained rules for "well-formed functions" (wff's), i.e., had a processer that could identify sentences. Once the system identifies a wff, it also creates new wff's by going to memory and retrieving other wff's, and combining them with the new one by means of logical connectives, "and," "or," "implies," etc.

According to the rules of the processer, every wff must be true or false. The inquirer sets itself the task of determining the truth or falsity of any identifiable wff.

As we have seen, the first step is to determine whether the wff is a tautology or a self-contradiction. The inquirer processes the sentence through the stored rules of logic. As I have said, there may be some error here, because the processer may not be able to test any given sentence with complete accuracy, and thus may fail to find a proof when one is possible. Furthermore, as we have seen, if the sentences are in a natural language, the problem of mapping them into a formal language to test for tautology may not be simple. Indeed, there has grown a sizable philosophical literature dealing with the subtleties of this problem (i.e., when is a sentence "analytic" and when "synthetic").

A sentence that is not identified as a tautology or self-contradiction can be called a "candidate." The inquirer takes the candidate and scans its memory for sentences that either imply the candidate or are implied by it. That is, the inquirer seeks to find a sentence p in memory such that "p implies q" or "q implies p" is a stored tautology, q being the candidate. If such a sentence p exists, and p is not identical to q, or a tautology or self-contradictory, then the candidate q is called a "contingent truth." In other words, a candidate sentence becomes a "contingent truth" if it can be linked to some sentence in memory; strictly speaking, it becomes a "contingent truth" if it can be linked to some prior contingent truth. Similarly, the inquirer scans memory to find a sentence r such that "r implies not-q" or "not-q implies r" is a stored tautology. If such a sentence r exists and is not identical to q, then not-q is contingent truth. Or, strictly speaking, not-q is a contingent truth if it can be linked by implication to a contingent truth.[1] Evidently nothing prevents both q and not-q from becoming contingent truths, and indeed this is commonplace in human inquirers. I may consider the candidate

[1] It may be noted in passing that the set of contingent truths contains no sentences which are false or true, and hence no contingent sentence implies or is implied by every member of the set necessarily.

sentence, "It will rain today." In memory I have stored the sentence, "It always rains on March 5 and this is March 5," and the sentence itself may have passed the test of contingency. This sentence implies the candidate, which then becomes a contingent truth. I may also have stored in memory as a contingent truth the sentence, "The newspaper predicts it will not rain and the newspaper is always right." This sentence implies the falsity of the candidate, which is therefore also a contingent truth. A candidate sentence may fail to become a contingent truth, in which case it is stored in a kind of limbo of irrelevancy, until such time, if ever, when it may be used.

The contingent truths belong to "fact nets." They imply other contingent truths or are implied by them. As the perception stream continues, these nets may grow. The very large nets are of chief interest to the inquiring system. As the net grows, certain sentences will become very critical, in the sense that if they are false, the entire net, or a significant portion of it, becomes false. Relative to the net, these sentences take on a privileged character: they become likely truths. It will be apparent that the tautologies are simply the privileged sentences of every net: if they are false, so is everything else. This suggests that the privileged contingent truths come near the "bottom" of the implication nets, i.e., they are sentences that are implied by most of the sentences of a net (a tautology is implied by all sentences in so-called standard logics).

Problem of Uniqueness in Leibniz's Inquirer

Even if the number of distinct symbols in the perception stream is finite, the number of sentences is (countably) infinite if the usual rules of sentence formation hold. If the inquirer's executive, using imagination, can produce wff's of any sort it wants, then its fantasy can create larger and larger nets of sentences. Without some sort of control, the Leibnizian inquirer would become a creator of nets of almost any kind, of any size. Such nets might be called "stories of the world," i.e., *Weltanschauungen*. As we shall see later on, such stories do play a central role in the inquiring system. But for the present it looks as though the inquiring system would get nowhere at all because for each story it can construct a counterstory with sentences that contradict the original story. The result is rather dull, something like an argument between husband

and wife, or a United Nations debate: "You've got it all wrong, that's not what happened at all."

In the area of the tautologies, there is also a multiplicity of possibilities because of the many possible dictionaries of terms: a sentence may be recognized as a tautology because what it asserts follows from a definition ("some blackbirds are birds" or "2 + 1 = 3"). If nothing controls the manner in which the inquirer constructs its dictionaries, there will be many alternative sets of tautologies.

In other words, the Leibnizian inquirer needs an executive who will exert control on the senseless proliferation of either contingent truth nets or sets of tautologies. Now one way to control the former is to insist that all sensory inputs can be regarded as contingent truths, and that any net containing the negation of a sensory input must be discarded. This choice was not made by Leibniz, and indeed the famous debate between Leibniz and Locke on the subject of innate ideas could be considered a debate about whether the optimal inquiring system has inputs from "outside" or not. Leibniz's inquirer does not, Locke's does. That is, Locke *seems* to insist that some outside influence will dictate which contingent truths should be taken most seriously, and it is not up to the executive to control this influence. Leibniz, on the other hand, wants the executive to do its own deciding internally.

In Leibniz's view the perception stream is all generated "from within." The inquirer starts with all of the symbols that it will ever need, and can construct all the sentences that are possible. The executive must work with this resource only, and must search the stream for just those sentences that form a coherent truth. This is the theory of "innate ideas" that was central to Leibnizian rationalism. However, what this theory means for the modern system designer needs to be explained later on at greater length. For the moment it is sufficient to point out that the inquirer is preequipped with a very rich classification scheme and can accurately place items in their proper classes.

It is easy to recognize that Leibniz's inquirer is essentially a model builder. The contingent nets it builds can be considered to follow an evolutionary process from embryonic models to full-fledged formal models. But what controls the model builder?

Leibniz's own answer consisted in finding one model that must dominate every other, because its own coherence implies its truth in an objective sense. The key concept is the familiar one in rationalism: God. God can be defined in such a manner that if the definition is consistent, such a God must exist. This is Leibniz's adaptation of the so-called ontological proof of God's existence: a thing defined to have all maximal properties must exist. The argument is a very subtle one. It consists of a careful analysis of "property" and the concept of the maximum of a property (e.g., "powerful" is a property, and "most powerful" is a meaningful adjective). It also requires a careful study of the consistency of definitions and especially the definition of a thing having all maximal properties—thus Leibniz's well-known statement that "God exists if He is possible." The last step in the argument is to show that there is one and only one possible model which includes the existence of a so-defined God. Hence, only those contingent truth nets that ultimately meet the requirement that God exists can be considered as candidates for validity. In Leibniz's philosophy, one and only one such net exists. In other words, the existence of God is sufficient for a unique solution of the system of reality.

Generalization of the Leibnizian Inquirer

Before examining Leibniz's problem of convergence to a unique model in more depth, it is useful to generalize on the Leibnizian inquirer. The very deep question of the role of God for system designers will be postponed for later chapters.

In generalizing on Leibniz's inquirer we want to maintain those features that are really essential, while relaxing on those that are more specific to Leibniz's own metaphysics. The list of essentials is as follows:

1. "Innate ideas" (i.e., no "inputs")
2. A capability of producing strings of symbols that break down into recognizable units (thus we drop the condition that the units be sentences, though from the point of view of inquiry this is one important kind of unit)
3. A capability of classifying any unit, e.g., a wff into either the class of tautologies or the class of non-tautologies
4. A capability of forming nets of units by means of a given set of relations and operators (thus it is not essential that the relation between the wffs be "implies," "and," "or," etc., and a Leibnizian inquirer can build nets between its units in various ways)
5. A capability of ranking the nets in terms of some prescribed criterion

6. A method of processing symbols and building nets, based on the ranking, such that the system will either eventually arrive at an optimal net and will know when it has arrived, or else will converge to an optimal net and will know that it is converging.

General Comments on Leibnizian Inquirers

We have already discussed the meaning of the first condition, namely, that all aspects of the symbol stream are under the control of the inquiring system. There is, of course, the more obvious meaning that the source of the symbol stream is "inside" the boundaries of the inquiring system. Curiously enough, this interpretation turns out to be quite sterile, even though some of Leibniz's writings indicate that this is what he had in mind (the monads have no windows). Many realists wish to depict the human mind in a harsh environment where outside stimuli impinge on the senses, much as a programmer impinges on a computing machine. Thus it is true that John Locke thought of the human learner as someone who receives inputs through his senses, presumably from an external world. But Berkeley clearly showed that the *designer* of empirical inquiring systems does not have to know where the inputs come from, and from his point of view they might just as well be thought of as internally rather than externally generated. It is really not until the time of Hegel that it occurred to philosophers that external vs. internal only makes sense to a third mind observing and/or controlling the process of learning. The three minds then become parts of the total inquiring system. Thus the component we have called the executive may find it convenient to recognize that one part receives inputs from another part, and is able therefore to differentiate between external and internal processes of these two parts. Specifically, those interested in communication between scientists will want to discuss the "output" and "input" in terms of publications, meetings, and the like. But in all such cases the whole inquiring system consists of the observer plus the two or more persons who are communicating, and hence there is no "externally" produced symbol stream.

The fact that the notion of an externally produced symbol stream is not useful to the designer does not imply a lack of interest in why certain symbols occur. But in Leibnizian inquirers, the explanation of the occurrence of a symbol event in the mind of the inquirer will itself be a set of symbols.

However, to be historically accurate it would not be correct to say that Leibniz's theory of innate ideas simply means that the Leibnizian executive has complete control of the stream and can make it do what he wants, whereas a Lockean executive must be buffeted by sensations willy-nilly. It is certainly not in the spirit of Leibniz's *Monadology* to say that the monads can create their own streams of perceptions; indeed, Leibniz's monads were all set in a preestablished harmony, so that their "reality streams" could no more be changed than could Locke's. In both systems, the executive has some latitude in using imagination to create new configurations of perceptions but is constrained by a "given" stream, so that the purported distinction between Leibniz and Locke does not hold.

The theory of innate ideas is much more closely tied into the sixth condition given above, i.e., the convergence on a unique model. Consider for the moment Plato's account of innate ideas in the *Theaetetus*. A boy who is not aware of the concepts of geometry is able to prove a theorem dealing with purely abstract geometrical ideas. He could not have gained this ability from his senses, and hence it must be innate. For example, he could not learn about straight lines from the senses, because he cannot see or touch a perfectly straight line. And yet he "knew" enough about straight lines to prove a theorem.

How does this idea of innateness translate? Clearly Plato's inquirer as well as Leibniz's have complete classification schemes built into them.[2] The problem of inquiry is to make these idealized schemes explicit. In Plato, the sensory stream "reminds" the inquirer of some feature of the classification scheme. In Leibniz, the ability of an inquiring system to prove the existence of God leads to a self-generated guarantee that the process of building nets of contingent truths will lead to an optimal net. In effect, the makings of an ideal net are built into the system and are not found outside it; the less than perfect inquirers must go through the cumbersome process of contingency nets in order to learn what their own nature is like. Thus the postulate of innate ideas of an inquiring system is the postulate that the system contains within its own structure the guarantee that the contingent nets converge on an optimal (condition 6).

The second function listed above permits the inquiring system to identify and individuate, i.e., to form units. It does not guarantee the

---

[2] For an interesting interpretation of this idea from a linguist's point of view, see Chomsky (1968).

simplicity or truth of some of these units, as Descartes required, because in Leibniz "truth" is an end point of a process, not a beginning (except for the logical tautologies and contradictions, and even these, as we have seen, are really "contingent" tautologies and contradictions). Leibniz devotes considerable time to the manner in which the inquirer identifies and individuates; he regards space and time as convenient devices that human inquirers use to overcome their inability to identify perfectly, i.e., by means of properties. This was an early recognition that the designer of an inquiring system must understand how the elements of the system are individuated, e.g., by some sort of an address system and identifying code, and that since the method of identifying and individuating is itself a design problem, one must search for criteria of effectiveness in this regard.

With respect to the third function—the logical tests—since Leibniz's day this has become a task of the executive. For Leibniz there was probably only one way to test whether a sentence is a tautology; today there are many, depending on the built-in logic of the inquirer: *Principia Mathematica,* modal logic, multi-valued logic, etc. In other words, the executive has choices as to which logic processer to use.

The fourth function—the establishment of nets—was based on the implication relation in Leibniz's system, and here it is generalized to any given set of relations. For example, it is generally recognized that "p implies q" and "p explains q" are different relationships; the nets could be formed out of explanatory chains as well as implication chains. Or the inquirer may introduce probability concepts, and consider the relation "q has maximum likelihood given p," where, as always, p and q may be conjunctions of sentences.

Certainly, the fifth and sixth conditions are the most difficult to design properly. Condition five—ranking of nets—requires some utility function applicable to formal systems. The most common choices are simplicity, elegance, economy of computation, and intelligibility, none of which is very precisely specified as yet in the literature, but see Goodman (1965) for a summary. In Leibniz's case, as we have seen, the quality of truthfulness was added; i.e., the model has some way of establishing its own validity. Even the more modest criterion of consistency of a net is not easy to work out in many cases, and may be impossible to establish in others. Much of the discussion of the next several chapters will be devoted to the problem of designing systems that approximate the fifth and sixth aims of the Leibnizian inquiring system.

Some Examples of Leibnizian Inquirers

Leibnizian inquirers represent one very general class of inquiring systems, and it is of interest to illustrate membership in this class by means of some recent work in artificial intelligence:

ALGORITHM MACHINES

These are machines that automatically derive solutions to certain classes of problems in a formal language, e.g., they find the maxima of a function, or solve simultaneous sets of equations, or solve certain types of differential equations. The machines store all the primitives of the formal language, plus the rules of sentence formation and inference. Hence they satisfy conditions 1, 2, and 3 given on page 34. The nets to be formed are strings of symbols that generate an implication net leading from what is given to a desired result (condition 4). Different nets can be evaluated in terms, say, of the computer time required to generate them (condition 5). There exists at least one finite net satisfying the requirements, and the system can check any net to determine whether it does indeed satisfy the requirements (condition 6). It will be noted that the executive is not completely free; certain instructions and information are given, i.e., are inputs.

HEURISTIC SEARCH MACHINES

Heuristic search machines, e.g., those using methods of "steepest ascent" to find the maximum of a function by a kind of "trial-and-error" method (Flood, 1964). These are Leibnizian inquirers, provided sufficient information is given them to guarantee the existence of a limit-point of the search. Here condition 6 may be satisfied in terms of an infinite convergence on the optimal.

THEOREM-PROVING AND PROBLEM-SOLVING MACHINES

The important point with respect to condition 6 is that the machine can always check to determine whether the theorem was indeed proved. It should also be noted that these machines may develop their strategies in terms of past experience, and hence they form an important subclass of Leibnizian inquirers. That is, the machine has an "executive" who discards one type of approach for another, if experience indicates that the former does not work well. It should also be mentioned that all existing problem solvers and theorem provers operate only on valid wff's, though not necessarily on relevant wff's. Some Leibnizian problem solvers can tell when a "datum" is probably false. Thus in Persson's "sequence extrapolator" (1966), the machine can guess that some of the data are erroneous, and can extrapolate the correct sequence from a subset of the information. In this narrow sense, the executive controls the symbol stream, which is therefore not a pure "input."

It is perhaps of some interest to note that Leibniz himself was explicit in declaring that the monads were not "machines"; see in Leibniz (1914), for example, his concept of automata in his discussions with Bayle. Leibniz seems convinced that automata cannot be "entelechies." The point is an important one, because it raises the question of whether "machines" can be subsumed under one or more of the following classes: (1) teleological entities (i.e., things to which a purpose can be assigned; (2) thinking entities; (3) conscious entities; and (4) entities capable of functioning through a value system. More needs to be said about each of these four questions, once we have used historical reflection to bring the problems of designing inquiring systems into sharper focus.

Leibnizian Inquirers and the Current Practice of Science

In addition to the machine designs mentioned above, mention should also be made of the fact that a great deal of the practice of science can be viewed as a Leibnizian inquirer. Indeed, after an excursion into the meaning of systems in the next chapter, we shall illustrate this claim in the field of organic chemistry. Despite the common notion that all observations of science have equal status so long as they are objectively obtained and have the same probable error, it is more accurate to say that a given group of scientists pays much more attention to a new result that can be linked into older findings, especially when the discipline is governed by theory. The theory provides the basis for tying together the results in the form of a "fact net." Thus a result that lies outside the largest net will often be ignored, whereas a result that enables the researchers to combine two hitherto unconnected nets will be acclaimed. Also, theoretical laws, like the conservation of energy, that come at the end of the net—i.e., the denial of which would entail a whole reconstruction of the net—are apt to be safeguarded by various devices. It is to be noted that the critical sentences lie at the bottom of the nets, i.e., are implied by many other sentences. Of course, the law of conservation of energy itself implies many things, but it is not this fact which provides its critical status, because it is possible to remove the *implicans* without necessarily disturbing the *implicatum*. But, as *modus tollens* tells us, to remove the *implicatum* necessitates removing the *implicans*. The fact net has hooks; if one pulls away at the top, the rest remains, but if one pulls at the bottom, the hooks catch and pull the rest of the net as well. Thus the sentence, "God established the orderliness of the universe," implies many things, but (according to today's science) is implied by very little: it can be removed from a fact net without disturbing the rest. But Euclid's Parallel Postulate could not be removed readily from his gigantic fact net, because it implied so much of it, e.g., that the sum of the angles of a triangle are equal to 180°, or that in a right triangle $a^2 + b^2 = c^2$ (of course, other sentences are needed to complete the implication).

It is often stated that one "counter instance" can destroy a theoretical assertion. Thus if theory T implies that A will be observed at place S at time T and non-A is observed, the notion is that T must be discarded. This strategy puts observational sentences at the bottom of the fact nets, and on the whole is a very poor strategy for a discipline to follow (as far as I am aware, no discipline uses it). A more subtle strategy consists of saying that T implies that A will be observed at S and T *by a perfect observer*. Hence the report of some observer, or even a group of observers, of the occurrence of non-A does not force the abandonment of T.

Finally, we may note that in order for a discipline to keep control of its nets and their growth, it makes some effort to exclude the relevance of results generated by other disciplines. Otherwise its fact nets are apt to be severely threatened. Whether all disciplines comply with condition 6—the need for a convergence on one optimal net—is of course not clear at all in an age that has turned its back on the quaint nineteenth-century notions of progress.

Leibnizian Concept of a Whole System

Finally, there are in Leibniz's design two ideas that need exploration in further depth, as we shall do in the next chapter. The first of these ideas occurs in all the rationalist inquirers: no optimal design of a part of a system is possible without prior knowledge of the "whole" system. To make this idea more precise we need to try our hand at defining "system" in a clearer manner. But the idea, if correct, challenges a great deal of the present world's philosophy of system design, for almost all our conscious policy making is based on the idea of fixing up messy situations wherever they occur. We "attack" poverty, inefficiency, national belligerence, crime, as though each were a blot on an otherwise pure white carpet, and as though we had no responsibility for showing how the whole system would improve if this part were changed in accordance with our plans.

The second idea seems peculiar to Leibniz: all systems are fundamentally alike in the design of their components. This idea that a "unit" of nature contains all of the complexity of nature is at least as old as Anaxagoras, who seems to have been the only pre-Socratic to have formulated it. Nor does it seem to occur with great frequency in the history of thought.[3] According to certain interpreters (Gershenson, 1964), Anaxagoras developed a structural theory of matter in which no matter how thinly nature is sliced in space and time, one always finds in any volume all the differentiation of kind that nature exhibits in the whole. Anaxagoras qualifies this remarkable stipulation by the remark that "in some things there is 'nous' also," which is variously interpreted to mean that some units have a "mind" or "intelligence" or simply a "force." Later history seems unanimously to have agreed that "some" in this quotation implies "not all."

Leibniz's "unit" is the monad, and everything that nature can express is potentially "in" the monad. Leibniz developed a teleological theory of reality, in which no unit of nature can function unless implicitly it contains all the richness of nature. To translate this into modern terminology, the idea is that in the design and evaluation of any functioning system the same set of considerations are always involved; to ignore any consideration is to design an incompletely functioning entity. The "considerations," for example, might be "sensors," "communication links," "feedback loops," etc. In this sense, all systems are alike, in that each contains all the complexity of nature, where "nature" is itself regarded as a designed system. The idea, in other words, is represented in modern system design in the search for the "general system."

The two ideas of rationalism, the need for an a priori theory of the whole, and the similarity of all systems, are important for system design and worth the separate consideration of the next chapter.

[3] Unless, of course, one includes all the various forms of Platonism; see next chapter.

: 3 :

ON WHOLE SYSTEMS:
THE ANATOMY OF GOAL SEEKING

The Anatomy of System Teleology

The reflections introduced by the historical excursion of the last chapter imply that the design of an inquiring system may well require an image of some "higher" system that provides the inquirer with its guarantees. In order to formulate this thesis in more precise terms, we need to discuss in greater detail the meaning of "system," and specifically the particular relationship that holds between a system and its parts.

We postulate that systems are examples of teleological things, i.e., things some of whose properties are functional. This postulate does exclude the "solar system" if we look at the planet and stars as modern science has taught us to do. It may even exclude "formal systems" like geometry in many cases. But if the exclusions seem improper, so that "system" is to be reserved for more general usage, then our postulate requests that we concentrate on a subclass for present purposes.

Indeed, the selection of a definition of "system" is a design choice, because throughout this essay it is the designer who is the chief figure. In other words, whether or not something is a system is regarded as a specific choice of the designer. We are to examine the ramifications of regarding inquiry as a system; other designers, e.g., the designers of experiments, may not regard inquiry in this manner. The ultimate answer to the question of whether inquiry, or the mind, or education is a system must be in terms of the effectiveness of the design strategy of conceptualizing each of these areas of human activity in this fashion. Therefore, what follows in this chapter is an explicit formulation of a design strategy, a commitment to the "systems approach."

The chapter is subtitled "the anatomy of goal seeking" to indicate the extraordinary complexity of the concept of a teleological system. Briefly, the necessary conditions that something S be conceived as a system are as follows:

1. S is teleological
2. S has a measure of performance
3. There exists a client whose interests (values) are served by S in such a manner that the higher the measure of performance, the better the interests are served, and more generally, the client is the standard of the measure of performance
4. S has teleological components which coproduce the measure of performance of S
5. S has an environment (defined either teleologically or ateleologically), which also coproduces the measure of performance of S
6. There exists a decision maker who—via his resources—can produce changes in the measures of performance of S's components and hence changes in the measure of performance of S
7. There exists a designer, who conceptualizes the nature of S in such a manner that the designer's concepts potentially produce actions in the decision maker, and hence changes in the measures of performance of S's components, and hence changes in the measure of performance of S
8. The designer's intention is to change S so as to maximize S's value to the client
9. S is "stable" with respect to the designer, in the sense that there is a built-in guarantee that the designer's intention is ultimately realizable.

Whether these nine necessary conditions are also sufficient is a basic question of this whole essay.

Teleological Classes

We turn now to a more detailed discussion of these conditions, beginning with the first, the concept of teleology. "Teleology" is to be defined within a (macro) cause-effect model of nature.

Since the details are somewhat tedious, a homely example may help to follow the logic. Consider an ordinary electric stove, with four knobs controlling the heat of the burners and pots of water on each burner.

The "world" to be observed is initially described in terms of the position of the knobs, the temperature of the burners, the state of the water in the pots (e.g., cool, warm, simmering, boiling), and a cook. Any two or more knobs in the same position belong to the same *morphological* class. Similarly, any two burners with temperatures in the same range (e.g., within 5° F. of each other) belong to the same *morphological* class, as do any two pots of water in the same state.

We regard the stove plus a cook as a sufficiently closed macro cause-effect system, in the sense that if we know the history of the system up to time t_0, we can predict the state at some later time t_1. We call the sufficiently closed system "macro" to indicate that our predictions are in terms of a morphological description; thus if burner number 1 has been turned on for two minutes, we can accurately predict the morphology of the water temperature in its pot.

Now suppose at time t_0, the cook turns one of the knobs but all else remains fixed. We would then say that the knob has changed its morphology. At a later moment of time t_1, one of the burners changes its temperature. We say the knob's position at t_0 *produced* the burner's temperature at t_1, meaning that had this knob been in any other position, some other temperature of this burner would have occurred. Also, at some still later time t_2, the knob's position at t_0 and the temperature of the burner at t_1 produce boiling water in a pot. We also note that the boiling water could have been produced by any other positions of the knob at t_0. In this case, we call the set of boiling-producing positions a *functional* class: the members of the class have a different morphology but a common product. Finally, we note that the cook can produce any turn of any knob. This means that he can produce functional entities. We call such a cook a purposive individual; the final end (the boiling water) is his purpose; and the set of things he can produce is a teleological class.

We now try to explicate the meaning of teleological classes more precisely by turning to a more formal and technical exposition of the ideas contained in this example. Singer (1959) provided us with a deep insight into the meaning of teleology. Functional classes, he says, are made up of entities that are alike with respect to their production of a certain end-result. More precisely, functional classes can only be defined in the framework of a cause-effect model in which aspects of any time-slice can be individuated and identified. That is, the designer must be able to construct a predictive model that tells him what events would happen if certain changes occurred in the relevant states of the world. An entity A, in a specific region of any time-slice t_0, is a producer of an entity B in another time-slice t_1 if it satisfies three conditions, which amounts to saying that t_0 precedes t_1, that A must occur in its region in t_0 if B occurs in its region in t_1 (i.e., if B does not occur in its region, the A does *not* occur in its region) and that A and B are proper subsets of t_0 and t_1 respectively.

It is clear in many cause-effect models that entities of quite different morphology can produce the same kind of an end-product. For example, many computers, once they are running, are virtually cause-effect systems. Some items in the memory may be essential for some later result, so that Singer's conditions are all satisfied (e.g., if the item does not occur in its special region at time t_0, the desired result will fail to occur at t_1). The item is therefore a producer of the result. But so are many other items in the computer's memory. Now if one concentrates on the producer-product relationship and forgets about the differences in structure of the items, then one thinks of the items as members of functional classes. For example, all items in S_0 that are essential for an output in S_1 belong to the same functional class.

Singer shows how one may effectively relax the conditions just specified for the definition of functional classes and thereby speak of entities having a common *potential* product. This is done by confirming, by the rules of the model, that some items of a given structure have produced a specific product in a given type of environment, and some have not. In this case, all items of the class are said to be potential producers. Further, one can introduce a metric in the classes and describe the probability of production.

Functional classes are more general than the specific type just defined. For example, one may want to define a functional class (e.g., a machine's output) in terms of things that are common products of one type of producer, rather than common producers of one type of product.

The extension of the definition of *function* to teleology is fairly straightforward. Suppose that some of the elements of a functional class that could occur in an environment are the output of one individual. If so, we call the potential outputs "means" and the common potential product the "end." The special elements of the functional class are then alternatives relative to an end-product. In this case we can call the class a teleological class and the end-product the purpose. Since all members of a teleological class are potential products of one individual, they are functional in two ways: as common potential products of the same producer and common potential producers of the same product.

Next, we note that we need not restrict our attention to one potential end-product. If we consider several, then the teleological class is defined in terms of several ends. But we also want to maintain the same sort of a metric that was developed for a single end. In other words, we want all teleological classes to be ordered in terms of a "more effective" relationship. Perhaps the simplest way to accomplish this is to weigh the end-products and develop a metric from the product of the weights and probabilities. The important point in weighting the objectives, however, is that the weights be functions of some property of the individual who can produce the alternative means. Specifically, we shall want to say that the weights correspond to the individual's "intentions" or "utilities" or "values." In this case, we can speak of the individual as a purposive entity.

In sum, a teleological class is a set whose members have a common producer and a common potential class of products, and can be ordered by a relation "is more effective than," which depends on the properties of the common producer as well as the probabilities of production. Thus an entity is teleological by virtue of the fact that other entities could (or do) exist that might produce the same results and might be produced by the same designer via his interaction with the decision maker.

We postulate that systems are teleological entities. This means that if the designer chooses to regard something as a system, he must construct alternative systems; he must then define the decision maker whose intentions are expressed in terms of the common potential products of the set of alternative systems.

Singer's account of teleology will receive more detailed consideration in a later chapter. As can be seen, its great advantage lies in linking together physical description with teleological description. There are, in fact, at least four ways of explaining the events of nature. First, the designer can explain one state of the whole system in terms of another (prior or past) state. This mode Singer called a cause-effect explanation. Second, the designer can explain how one aspect of the whole system influenced or was influenced by another aspect by virtue of its physical or morphological properties. Singer calls this a producer-product explanation. Third, the designer can explain how one aspect influences or is influenced by another by virtue of its functional properties. Finally, the designer can explain events in terms of purposive behavior, i.e., the choice of functional entities.

For more detailed accounts of teleology, see Churchman (1961).

The Client, Decision Maker and Designer

In the description of teleological systems just given, it is essential that there be a purposive individual who can produce alternatives that lead with varying degrees of success to his desired objectives. But we can distinguish three such individuals: the client, the decision maker, and the designer. The rather intricate interrelationships between this cast of characters need to be explained.

The client can be described in terms of his value structure. For him there are a set of possible futures, i.e., states of nature, and he has a real preference for one state over others. More specifically, we can describe the possible futures in terms of a set of properties, which we call "objectives" or "goals," and the client's interest in each of these properties can *in principle* be described by a "trade-off" principle that tells us how much of one objective he would be willing to relinquish in order to increase his amount of another objective. In order to estimate this trade-off policy, the designer imagines a world in which the client could change things as he wishes, within the bounds of limited resources (e.g., limited amounts of money). Since his resources are limited, he cannot create his ideal future, but instead must create a future which comes as near to what he wishes as his resources will allow. The designer seeks to find the underlying principle behind the client's trade-offs by estimating a "measure of performance" which enables him to assign numerical values to possible futures. In our case, a "possible future" is a specific state of a system and the products it can produce. The designer is successful to the extent that he can accurately measure the client's real preferences. Many examples of this effort are to be found in the literature; see Ackoff (1957), Fishburn (1964), Ackoff (1961).

Some readers will feel considerable doubt about the success of the implied program of the preceding paragraph, and we shall want to explore these doubts later on. The doubts, in fact, relate to item 9 in the list on page 43. For the present, however, we assume that an estimate of the client's intentions can be obtained in the form required by item 2, namely, a measure of performance of S.

It is to be noted that the designer's intentions are always "good" with respect to the client; that is, the designer's value structure is identical to the client's. But the third character, the decision maker, has a somewhat different relation. It is he who controls the resources and hence creates the real future. More precisely, the decision maker coproduces the future along with the environment, which he does not control. In the stipulations listed above for the existence of a system, nothing is said about the good intentions of the decision maker. He also has a value structure, i.e., a trade-off policy for alternative futures, but his trade-off principle need not be identical to the client's and designer's.

Thus we can recognize the tragicomedy of the designer's conceptualization of the real system. He sees a client whose interests he believes should be served by the system and a decision maker who produces the futures. If the decision maker's intentions are not "good," the situation tends to become tragic. The designer's role will be to try to change the decision maker, i.e., to change his value structure. The complexity of the problem is not in the least bit eased by the reflection that all three characters may reside in one person. For example, the question still remains whether such a person should choose himself as the client of the system he wishes to design.

It becomes clear that one of the designer's most difficult problems is to identify the client and the decision maker. Although I have referred to each as though he were a single person, this is obviously a fiction. In reality, both client and decision maker are highly complex entities, made up of interacting forces. But perhaps more important is the designer's choice of the client, i.e., the complex of persons whose interests *ought* to be served. The simplest basis of choice is to say that the client is the group of people who pay the designer for his work, subject to the obvious constraint that the client's intentions are legal and, perhaps, moral in some broad sense. That is, the designer is moral if he serves a client who has a legal or moral right to expect that the system will serve his (the client's) interests and his interests are themselves legal or moral.

This design strategy in the selection of a client seems to be the most practical one, and is widely adopted today by designers (engineers, management consultants, architects, planners). However, there are very strong traditions that argue for a much deeper analysis in the selection of the client, one in which the long-range implications of the system are spelled out. The extreme of this position is that there is one right way for systems to go in terms of man's "ultimate" goals, and hence that the designer should select that client who understands best the ethical basis of long-range planning. Thus we can form a classification of design strategies that are relevant to this aspect of the design process:

1. Design (of systems) is appropriate for short-range goals, but

not for long-range "ultimate" objectives. Hence choose the client whose short-run aims are legal and who can legally expect the system to serve these aims. Such a strategy is teleological in the short range, ateleological in the long range.

2. Design is appropriate for both short- and long-range goals, the short-range serving the long-range "ultimate" or "ethical" goals. Hence only choose the client whose long-run aims are estimated to be ethical. This strategy is teleological in both short- and long-range goals.

3. Design is appropriate for long-range goals, but not short-range. The appropriate short-range decision is based solely on the moral quality of the act (e.g., its honesty), and not on what the act produces in terms of goals. Hence, choose only moral clients (i.e., clients who act from moral motives alone), and do not design short-range systems for them, but rather help them to discern clearly the moral issues of their choices. On the other hand, the designer can help the client to see the long-range objectives of the moral life, where the "system" emerges, in Kant's terms (1788), as a "kingdom of ends" where happiness (maximized benefit) and virtue coincide. Hence, the systems approach is appropriate with respect to man's ultimate goals, but not his immediate goals; i.e., the strategy is teleological in the long range, ateleological in the short range.

4. Design of systems is inappropriate for short- and long-range goals. That is, do not choose any clients, because what you as a designer do is inappropriate (immoral, ugly, meaningless, etc.). This strategy is the "deadly enemy" of design; it is "anti-planning." We shall keep it in the dressing room until the last act. Our concern for the time being will be mainly with the second option, and especially its relationship to the first.

Teleological Components

The preceding discussion has concerned itself with items 1, 2, 3, 6, 7, and 8 of the list of specifications for a system on page 43. We turn now to item 4, which deals with the system's components.

Not all teleological entities are systems. The differentiating feature of systems is that they can be separated into parts and that the parts work together for the sake of the whole. The reason for this design choice seems more or less obvious. Most systems, e.g., human bodies or human organizations, are difficult to understand as unified wholes. By breaking them down into components, the designer may be able to formulate the necessary details. The parts are themselves teleological, i.e., can be regarded as potential producers of a set of products (goals). What is of chief interest to the designer is the relationship of the parts to the whole system.

Now a rather obvious way to express this relationship is to say that the parts "make up" the whole system, and even to insist that a part must "belong" to the system in the set-theory sense of belonging. However, this very strict stipulation does not seem to capture the spirit of design. What matters most to the designer is that he can conceptualize how a decision maker can change (design) a part and the change makes a difference in the performance of the whole. To make this idea more explicit, suppose the designer has an approximate measure of the whole system's performance over a period of time. Call this measure M. He now seeks to find a measure of performance of each component m_1, m_2, etc., which will have a certain relationship to M. When we say that the designer can change a component, we mean that he (through the decision maker) can change its measure of performance somewhat. From the designer's point of view, there are at least three types of relationship between the component measures and the total system measure:

Weakest (necessary condition): M is maximized (i.e., the whole system performs perfectly) if and only if every component measure, m_i, is maximum;

Moderate: A positive change in the value of m_i is a producer (or probable producer) of a positive change in M for at least some range of values of m_i. (This is stronger than saying that M is positively correlated with m_i; it is equivalent to saying that M increases when m_i increases in the range, all other states of the system being held fixed, and M would not increase if the part were changed to some other morphological or functional class);

Strongest: There exists a mathematical formula which expresses M as a function of the m_i's only; and the "global" maximum of this function exists.

Thus, to summarize, S is a system to the designer only if: (1) S is regarded to be teleological and hence to have a measure of performance M; (2) S is regarded to have teleological components each with a measure of performance m_i; and (3) the designer can conceptualize how changes in the components' measures of performance m_i produce changes in M.

In symbolic terms, we imagine a class S of alternative whole-system designs. S is mapped in a one-to-one correspondence onto a bounded real number scale, so that each M_i of the real number interval represents the measure of performance of each S_i of S. There exist sets of alternative component systems s_1, s_2, etc., with similar measures of performance m_{1j}, m_{2k}, etc. Any M_i of an element S_i of S, in the strongest sense of "system," is a monotonic function of the measures of performance of the components.

Some remarks on the relationship of the measures of performance are in order.

1. The three ways of relating M to the m_i of the system components express the attitude of the designer. In the weakest case he does not wish to commit himself to any explicit relationship; he hopes by some means, e.g., "trial and error," to find a way of designing the components which will maximize the performance of each one, and hence maximize M. In the moderate case, he hopes to improve the system part-by-part. If, however, the process of improving a part's performance begins to interfere with the performance of other parts, he turns his attention elsewhere, again on a trial-and-error basis. This design tactic is quite common in medicine and social policy formation. The doctor seeks to cut down on cholesterol in order to "improve" the blood circulation system, but in the process the patient gains weight; the prescription is then changed to control diet. Or a city tries to improve housing, only to find that the people of the ghettos become socially aware of their state and more prone to revolt, since a certain degree of affluence is necessary to initiate revolution. In the strongest method, the designer is willing to commit himself to an explicit functional relationship ("model") between M and the m_i, so that in principle he can estimate the optimal design of each component by mathematical means.

2. In any choice the designer makes, he may identify the components incorrectly and/or measure their effectiveness incorrectly. Thus he may believe that he understands how M is related to the m_i, and hence make the strongest approach, but he may be wrong. This rather obvious point merely states that there is a real relationship between both the true M and the true m_i, and the designer's estimates of these. The strategy the designer uses to study the system (weakest, moderate, or strongest) need not be evidence of this real relationship.

3. Finally, the relationship of the components to the larger system brings in number 5, the environment, in the list of specifications of a system on page 43. The environment is not controlled by the decision maker, even though it does coproduce the relevant states of S and hence S's measure of performance. In the definition of producer-product given

above, all producers require an environment in order to produce (i.e., the producer is not a complete cause of an event). Hence, even in the strong sense where M is a function of the m_i, the nature of this function depends on the (uncontrolled) environment, and varies if the environment changes over time. From the designer's point of view, what is environment (changes not produced by the decision maker) and what is component (changes produced by the decision maker) depend on who the decision maker is. Since the client's interests are the same as the designer's, the designer should choose as the decision maker that complex of wills (potential producers) which the designer's conceptualizations can influence to produce the maximum gain in S's measure of performance relative to the client. Hence the designer should always conceptualize how his recommendations can be implemented in order to select the optimal decision maker. It must be admitted that in much of urban and industrial planning, the planners tend to ignore this piece of common sense.

These last remarks show how important the designer himself must be to the task of defining a system. Indeed, the designer needs to have a theory of his own role as well as a theory about the system. He must try to understand how he can learn about the system, what influence he can have on the system's changes, why he should exert influence, and so on.

Although this account of a system may seem more or less straightforward, there are a number of practical difficulties implicit in its meaning. Most federal government agencies must yearly prepare a budget which is to be approved by the Bureau of the Budget and congressional appropriations committees. Suppose an agency can modify the behavior of these review organizations. One can see that if the agency is the designer, this fact may imply that the Bureau of the Budget or the appropriations committees are "parts" of the agency viewed as a total system. Of course, the designer may want merely to say that those in the agency who present the agency's case for the budget proposal are parts, the attitudes of the people in the Bureau or Congress being fixed and beyond the agency's capability of changing.

In any event, one can begin to sense the delicacy of planning for change of whole systems. The pathway to the optimal whole system may not consist of improving each part, one by one, until it is perfect, because once one part is perfected and then another part is changed, the first may lose its top position. For example, in an automobile the wheels may be designed so as to give optimal performance for a given engine; but the "best wheel" may become the "worst wheel" if the power of the engine is increased. I know of one very "successful" study of a company that maximized the performance of its marketing department; the only trouble was that the product was inferior to many others on the market. Hence the designers had "perfected" a way of selling a bad product.

In other words, the rank order for one part of a system may completely change if another part is changed: the rank orders may be functions of the state of the other parts. We shall want to consider two kinds of systems, one in which the rank orders remain fixed independent of the other parts, and the other in which the rank orders depend on the state of the other parts. The designer's role may be simplified if he can concentrate on one part at a time and try to move this part to a better level of performance before going on to other parts.

Separable Systems

In such systems the parts are "separable," so that the effectiveness of one part is virtually independent of the state of the other parts of the system. More precisely, we say that a system S is *separable* with respect to some part if the measure of effectiveness of the part is independent of the states of the other parts. This means that whatever happens to the other parts, this part's contribution remains invariant.

As an example of a non-separable system, consider a factory. One typical way of partitioning the system is to break it into these subsystems: procurement, order processing, production scheduling, production control, labor force, inspection, packaging, and distribution. In most production systems, none of these parts is separable. The optimal procurement policy always depends on the way in which items are scheduled, the optimal production scheduling depends on how labor is deployed, all parts depend on the extent and timing of control, and so forth. Indeed, most students of production question whether manufacturing itself is a separable part of the whole organization (e.g., the optimal manufacturing policies depend on the state of the subsystems that control investments or prices).

Nevertheless, in an attempt to simplify, the designer often tries to design a production scheduling system by more or less ignoring the other aspects of the firm. As a result he may try to consider the system as virtually separable, recognizing some risks in doing so. For example, if the development of labor changes, or the marketing system is modified, the designed production system may then become worse than it was before it was redesigned.

On the other hand, consider two men assigned to dig a trench. If each man is considered to be a part, then the system might be considered as a separable system: the work each man accomplishes adds to the total independently of what the other does. Some systems that collect and store items may also be construed as separable. Also, a system designed to solve problems in a formal framework may be separable: the optimal method of solution of one problem may not depend on how the other problems are solved.

It is not difficult to describe the concept of separability in precise terms, as is done in systems engineering. Suppose there are two parts, one controlling a variable quantity x, the other a variable quantity y. Suppose that M, the system performance, can be expressed as a function of x and y, $E(x,y)$. Suppose $E(x,y)$ is a continuous function with partial derivatives E_x and E_y. Finally, suppose that E_x is a function of x only and E_y is a function of y only; i.e., $E(x,y)$ can be expressed as $A(x) + B(y)$, where $A(x)$ is a function of x only, and $B(y)$ is a function of y only. Then the system is completely separable.

Design Separability

However, this very common way of defining separability is not appropriate in the context of this book because we are concerned with the designer and not some abstract system that exists independently of the designer. Consider, for example, the rather "obvious" separability of the ditch-digging system. How does the designer know that the parts are separable? Indeed, as soon as we raise the question we see that even in such a simple case it is not at all obvious that separability obtains: perhaps the manner in which one man works could enhance the other's task. The designer's question is: What do I have to know about the other parts in order to know that this part is separable? Thus we can easily imagine a case where a part is indeed separable, but where a designer could not possibly acquire information on this point without evidence about all the parts.

If we return to the industrial example, the problem of the designer can be made clearer. When a designer examines a procurement policy, for example, he normally attempts to measure some of the relevant properties of the system: the way in which demands come into the system, the way in which prices vary, the costs of placing orders, the delays in receiving orders, and the costs of storing items. These measures are then combined by a model to form a measure of procurement effectiveness. In attempting to measure these properties, the designer must examine how other parts of the whole system are operating. The demand, for example, is the consequence of policies of the production department or sales department. The cost of placing orders is the consequence of policies of the order department. The cost of holding stored items is partially the consequence of the firm's investment policies. Now these policies of other parts of the system can be varied; in fact, they determine the effectiveness of the part with which they are associated. As a consequence, any so-called optimal policy of the procurement system is asserted to be optimal only because one assumes that the relevant policies of the other parts are optimal.

In other words, the designer has no way of obtaining information about the effectiveness of a part except to assume certain characteristics of other parts. The parts, therefore, are not separable in a design sense.

The concept of design separability of the components of a system can now be formalized by extension of the language introduced above:

1. A "choice set" C relative to an individual is a set of potential choices of the individual.
2. An individual "knows about" a choice set C relative to an objective O over a set of environmental conditions N if he chooses the optimal member of C in all conditions. Normally, we add the stipulation that the conditions so change that the optimal member of C changes, and hence that the choice of the optimal is invariant with changes in the structure of the optimal.
3. In general, one individual A "knows more about" a choice set C than another individual B if A is "more likely" to make a better choice over a set of environmental conditions N than is B (the definition of "more likely" need not divert us here).
4. An entity x is "evidence" relative to an individual, a choice set C, an objective, and a set of conditions, if x makes the individual more knowledgeable.
5. A designer is a system S_1 with the objective of improving a system S_2.
6. One part s_{21} of S_2 is design-separable from another part s_{22}

relative to a designer S_1 if selecting some aspect of the behavior of s_{21} as evidence about s_{22} for S_1 is never an optimal strategy of S_1 relative to the objective of improving S_2.

Various types and degrees of design separability could now be introduced, e.g., the amount of difference in improvement that evidence about another part produces, whether the design separability depends on that state of S_2, and so on.

It is important to emphasize that in introducing the concept of design separability, the concept of separability is made relative not only to the system and its parts, but also to the state of the designer. Thus it may be that the effectiveness of a system can be represented, as suggested above, in the form $E(x,y) = A(x) + B(y)$, but a designer may not know this, and the only way he could come to learn it would be to know all about both parts.

The concept of the designer of a system also helps to shed some light on a question raised in Chapter 1 and one that is central to Chapter 2: the "size" of a system. From the point of view of one designer, the size of a dwelling system may simply be a house or a room in it; from another designer's viewpoint it may be the community or nation. This suggests that the "size" of a system is defined by the choice set of the designer: anything the designer can change through the decision maker which will influence the effectiveness of the system is a "part" of the system. On the other hand, information about some aspect of nature may be very important in designing a system, even though the designer cannot change this aspect. But "learning about an aspect of nature" is an activity of the design system, and it is an activity that may be subject to the choices of the designer. Thus the design system may include "learning about x" as a part, even though x is not a part of the system being designed.

Illustrations

We can now illustrate the system specifications 1 to 8 on page 43. First, consider a college in the following manner:

1. The college has a set of goals
2. The measure of performance is the number of student credit hours per dollar of expenditure per semester

3. The client is the set of people who can potentially attend the college plus those who pay the tuition
4. The components are the curricula of the college, plus administration and services
5. The environment consists of legal constraints, budgetary policy, community reaction to the college, etc.
6. The decision maker is composed of trustees, administration, and faculty
7. The designer is (say) the planning committee of the college
8. The planning committee's intention is to recommend changes in the college programs that will maximize the college's benefit to the client.

Evidently, there is much to criticize in this "system's approach" to a college, even though it closely approximates the way in which many trustees and state legislators regard these institutions. The weakest point perhaps is the measure of performance, because this can be increased simply by making classes larger, which clearly does not serve the client's interests. What is not evident at all is a more adequate measure. This is a point to which we'll return later in this book, when we discuss the relation of individual value to system value; specifically, we'll want to consider the thesis that the purpose of a college is to foster every student's unique style of learning. Another decidedly weak point is item 4, because it may very well be that the existing divisions into disciplines and curricula fail to capture the spirit of the learning process. Finally, item number 6, though essentially valid for most colleges, may represent the real weakness of today's policy of higher education, because there is much to be said for the thesis that the student is an "expert" on the learning process and should be a prominent part of the decision maker. Again we'll want to review this argument in greater depth when we discuss the limitations of the systems approach.

As a second illustration, suppose we consider "basic research," a much discussed but little understood area of human endeavor. If we regard basic research as a system, then the following suggestions emerge:

1. Basic research has a set of goals
2. The measure of performance is the amount of "new and significant" knowledge produced per unit of research effort
3. The client consists of all persons with intellectual curiosity
4. The components are the scientific disciplines; within each dis-

cipline the subcomponents are individual researchers or research teams (we shall consider the organization of these subcomponents below)

5. The environment consists of legal, budgetary, and social constraints, plus educational policy dictated by trustees, administrators, foundations, government agencies, and legislators
6. The decision maker is the community of scientists
7. The designer is the community of scientists
8. The community of scientists "plan" for the best development of basic research.

One can sense how introverted this system becomes, especially when the designers suggest that the rather vague measure of performance (item 2) be the result of the subjective judgment of the scientist. Hence, what research is supported (within the budgetary constraints defined by the environment) is decided by the community of scientists, who, together with their admirers, are also the client and decision maker. The paradox is that basic research, which is supposedly the most objective method of learning nature's secrets, becomes completely subjective when viewed as a system: as a system, its performance measure is based on a collective subjectivity not checked by external judgment. From the system's point of view, it does not differ from an esoteric mysticism which sets its own criterion of inner truth. To break out of this shell, some designers suggest we judge the value of basic research in terms of its "payoff" in applied technology. But to others in the scientific community, this suggestion seems to rob basic research of its "basic" property. The problem is a very delicate one: How can we at one and the same time design a system of basic research which preserves the traditional freedom of inquiry and yet becomes more objective in its value structure?

The question is closely related to the question of the boundaries of the basic science system. The description given above assumes that the system is made up of the disciplines. But should we not include *within* basic research those efforts designed to procure funds for research? Are the policies of funding really in the environment and hence beyond the control of the scientists? The policies of funding are in part determined by politics.

Now internally science has a politics of its own, as the disciplines debate their respective cases. But "science" also attempts to influence business and governmental policies. We sometimes tend to think that the politics of science is good because, I suppose, everyone wants more knowledge and therefore the political activities of those who acquire knowledge for man must be sound. More specifically, are we justified in saying that any activity which reduces our ignorance of the consequences of our policies is justified? If we view the problem as one of ignorance vs. knowledge, the answer must be affirmative. But if we view the problem as one of ignorance vs. deprivation of goods, then the answer may be negative. It may be better to be ignorant than to starve. A little more honesty may lead us to suspect that the kind of research that receives large grants may not be justified in the larger scope of national interest. Thus in the broader viewpoint one cannot distinguish between science and its politics; it makes no sense to the designer to say that science is a body of knowledge and politics is people, and therefore the two must be different. For the designer it is impossible to optimize the system of acquiring basic knowledge without considering the political problems that such a system generates. The boundaries of "basic research" expand into the area of national policy making, and the client becomes larger than the scientific community.

And there is some good ground for thinking that information about the government and international politics is relevant to the design of specific research projects, once we realize that science is an institution devoted to survival. Its programs are designed for long-range objectives that cannot by their nature be accomplished in a generation. Hence, one strategic problem of a scientific project is the optimal bequeathing of the project to others, just as one overall strategy of science is bequeathing its work to the next generation.

If we admit the bequeathing problem to be an essential part of the problem of scientific method, then we must include within the design system of science the strategies of its survival. The consequence is that the design system of science will become very large in eras when politicians and industrial managers seek to use science for ends that conflict with the aim of scientific control. Hence, the design system of science ought to include estimates of optimal international policy, if international policy could threaten science's existence.

At this point we can afford to pause and ask whether the conclusion just reached makes a difference. Of course there are very many scientists who do not believe that world politics is a scientific problem. Their disbelief is based on their implicit assumption that "science" is simply the activity or output of some more or less well-defined group of researchers. What I have called the design system of science is therefore not conceived as a "part" of science itself. But most of these scientists would include the logic of scientific inference in the corpus of science. It follows that they are willing to include in the "body of knowledge" some aspects of the design system. If this much is included, then why not the rest? On what basis are we to draw the border line between the internal strategies of science and its external strategies?

Still, even if those who want to keep science pure are illogical, does the issue make a difference? Most scientists are concerned about world politics and would like to help. What difference does it make whether they regard the issues as scientific or nonscientific?

There are at least two ways in which it does make a difference. The first has to do with attitudes and the second with reflection. If communication and world politics are regarded to be scientific issues, then the status of those who work on these issues is improved in the scientific community. Much more important, scientists themselves will realize the necessity of rigorous analysis and controlled fact finding in the study of these problems, qualities that seem to be lacking in many of the current discussions. Finally, if these issues are taken to be scientific, then our best minds may want to weigh the scientific value of working on them rather than on matters of energy and space.

The second difference is reflective. I said above that most scientists would place the logic of inquiry in the corpus of science. But I think that very few of them have ever tried to check this logic in the same sense in which they check discoveries in their own fields. The validation of the logic of inquiry consists in showing that certain activities carried on within the scientist's laboratory or study are optimal ways of reaching results. How can the scientist *guarantee* this optimality?

Research Activity As a System

This last question is concerned with the optimal design of a research project (component 4, see page 56). Although scientists do speak of experimental "design," it is seldom if ever that the process of experimenting is viewed as a system. For example, there is the prevalent assumption that thinking and/or observation are separable parts of a research project.

This position is the essence of positivism. The crucial point is not that thinking or observation may fall into error; everyone recognizes this possibility. The positivist position is that it is possible to do the best one can with some acts of thinking or observing without having to be concerned about the uses to which one's thoughts and observations are put or the way in which they are communicated to others. But from a systems' point of view the question is whether the observational part of the system is actually design-separable from the rest of the system; Descartes and Leibniz claim that it is not, positivism that it is.

To consider this question in more detail, suppose we try a systems' approach to a research project by constructing a more or less obvious division of the research activity as follows:

Part 1 determines what specific objective the team shall pursue, i.e., defines the problem area;

Part 2 specifies the problem, i.e., creates a model within which the problem can be defined (the model often being expressed in a formal system);

Part 3 determines the logical consequences of (theorems) of the model;

Part 4 specifies what data are required, in what form, to what degree of precision, in what amount;

Part 5 specifies how the data are to be collected;

Part 6 collects the data according to the requirements;

Part 7 transmits the data to a central point;

Part 8 analyzes the data;

Part 9 produces a set of results;

Part 10 stores the results and transmits them when needed;

Part 11 determines when stored results are needed and how they are to be used.

The positivist thesis is that inquiring systems can be so designed that Part 6 is separable. Part 3, often called analysis, is also a likely candidate for separability.

We can readily satisfy ourselves that some of the other parts are not separable, and indeed that their proper design is still an unsolved problem. Consider, for example, Part 10, the part of research that stores and transmits information. In earlier times, men thought of communication systems in relatively simple terms, perhaps because the total amount of real information was small. The alternative designs consisted of mixtures of talking and writing, of personal memory and written documents. Today, we all realize that the problem of storing and retrieving information has become serious and that the number of alternative designs is very large.

If we were to start from the beginning, would we build libraries with books? Would we publish journals? Would we hold meetings with papers? Would each scientist's study be equipped with pushbutton panels that would call up what he needed to know on a television screen? What form would such information take? How would a researcher in such an environment know when to request information? At our present stage of technological development, we have some valid reasons for suspecting that science is poorly designed in its communications, so poorly that the resulting inefficiency may be colossal.

In any event, it seems clear that information-retrieval subsystems are not separable. For example, the optimal information system depends on how the results are to be used (part 11), as well as the form of the results (part 9) and their structure (parts 2 and 8). In other words, evidence about the state of the other parts is required to design part 10, and improvement in part 10 depends on the status of the other parts. Thus part 10 is not separable and hence not design-separable either.

The question raised earlier was whether the data-collecting component is separable. This would be the case if it were possible to observe and record observation accurately, without cost, and without depending upon the states of the other system components. When scientists speak of "raw data," they often have some such idea in mind: the raw data are received by the inquiring system in the ultimately simple manner. We shall be devoting some space in chapter 5 to the design of raw data-collecting components, and our conclusion will be that these components cannot be designed in a separable manner. But even without this analysis, it seems almost obvious that the requirements of accuracy and minimum cost, if legitimate, make the design of data collection nonseparable.

The theme that emerges from these discussions of systems is that all systems are design nonseparable.

Systems and Their Design Systems

This thesis takes us back to the issues raised at the end of the last chapter, as well as to the last item of the list of system specifications on page 43: what kind of guarantee must the designer of a system have that his activities are meaningful relative to the system objectives, given that evidence of the improvement of a part depends on a knowledge of the "whole" system?

For all three rationalists, Descartes, Spinoza, and Leibniz, the design system is an integral part of the system itself; furthermore, the system designer must assert that one of his fundamental activities is the development of system guarantors. Specifically, a fundamental problem of science is the problem of guaranteeing that the activity of inquiry by any specific scientist is meaningful relative to science's objectives. This does not imply that each and every scientist works on the guarantor problem; it does imply that each and every scientist has the responsibility of making sure that the problem is worked on.

It may seem odd to include design stability, i.e., guarantee of real improvement, into the definition of a system. But it should be emphasized again that we are interested here in the meaning of a system from the designer's point of view, i.e., in the "systems approach" to social change. If the designer's world contains no guarantee that his activities will improve the system for the client, then the real system is largely meaningless for him.

Now the most forceful reply to the challenge of rationalism is to argue that the rationalists ignored the possibility of viewing a system as adaptive; they assumed that the problem of the guarantor must be settled once and for all. Instead, shouldn't we recognize the importance of "temporary separability" of system functions, in order that we can get on with the job? That is, the designer temporarily assumes separability in order to improve a part, and later on may drop this assumption. The designer wants to proceed in this way because he feels that temporary separability is the only feasible approach to system design. It permits control relative to fairly specific objectives; it permits an adequate scanning of alternatives and a reasonable evaluation of each. Hopefully, then, if one can perform reasonably well in each segment, the whole will develop in a reasonably satisfactory manner. And if not, then the designer can introduce decision rules for revising the process of change, i.e., for adapting to unforeseen consequences. In the case of inquiry, the dominating criterion of control is objectivity; one wants to be reasonably sure that the evidence for a state of affairs is not itself distorted by the feelings of one of the investigators, or some external but unknown influence. There is overwhelming agreement that the larger the system, the greater the risk of non-objective evidence.

But the concept "temporary separability" does not avoid the guarantor problem; it means that the designer of a system assumes that the whole system has certain specific characteristics as far as its impact on a part is concerned. It is stable, or changing very slowly, and changes in the part will not influence its stability. He doesn't know that this assumption is correct, but for the time being he acts as though it were, in order to be able to grapple with a smaller problem by means of reliable and precise methods.

Some support of this approach seems to be found in our recent heritage in matters of inquiry. Modern mechanics began with the study of one body, and when this system was understood, it went on to the study of many bodies, of fields, and so on. So in modern game theory, we find students of conflict who believe that one starts with a simple constant-sum two-person game, and that the optimal behavior of people in such a situation can be fully understood regardless of any other conflict situation these people may be involved in, and regardless of any other characteristic of the world.

Thus, even though separability may never occur in a pure and permanent form, shouldn't we act as though it holds, insofar as we can legitimately do so? The principle that motivates one to answer this question affirmatively may be stated as follows: design the inquiring system in such a way that some of its parts are *virtually* separated. Proceed in the study of the separated parts and only reconstitute the system when separability is no longer feasible. This is surely the spirit of academic research and a large part of business and government research.

The opposition between classical rationalism and the contemporary theory of system development needs to be stated more precisely in order to bring out the essential characteristics of the rationalist inquiring system.

Dynamic Design Systems

Consider, as before, a system S with subparts s_i. Instead of considering S as a fixed entity in time, consider a method of designing S whereby the designer takes the parts to be separable so long as the system "behaves properly," and redesigns the parts whenever the system fails to operate properly. We shall call the principle by which a part is changed a "transformation function." The crucial point in the design is whether one can recognize the unsatisfactory state of a part without having to study the entire system in depth. This is equivalent to asking whether the transformation functions are functions of the prior states of S, or merely functions of a subclass of the parts. If they are functions of S, then the designer needs to know or estimate the properties of the whole system in order to judge how to change a part.

The strongest form of the separability principle with which we are concerned states that the transformation functions of a part are functions of the prior states of the part only. Specifically, it says that a designer can partition S into subparts in such a manner that:

1. Decision rules can be constructed for the operation of each part
2. The application of the decision rules to a part depends only on information about prior states of the part
3. There exists a function of the prior states of a part only, which determines whether the part is stable (satisfactory) or unstable (unsatisfactory)
4. There exists a decision rule applicable to the system that will so modify its operating rules that the part will be transformed from an unstable part into a stable part
5. As soon as a certain degree of instability of a part occurs, the designer can recognize this event and influence the decision maker to change the system so that the part becomes stable
6. The set of all possible decision rules governing each part, plus the transformation rules that send a part from an unstable to a stable state, according to 1 through 5, contains at least one member that is superior to any other rules for operating S, relative to the client's objectives.

In summary, the designer keeps his eye on the various segments of the system, and when he detects serious trouble, he moves to modify the segment. The approach is "incremental"; one moves in steps that the decision makers and clients can understand.

The reader may recognize that this principle of "incrementalism" underlies a good deal of present-day reflection about adaptive behavior. See Lindblom and Braybrooke (1963). For example, the principle is inherent in statistical quality-control procedures. Inspection, which plays the role of an inquiring system for production, partitions the production system into parts, identifies the properties of the parts, sets up standards of stability in terms of control parts, signals instability when it occurs solely on the basis of the data obtained from the part, and reconstitutes its image of the system until the "assignable" cause is found. Similarly, the model of a "satisficing man" shows him to be one who breaks out reasonably sized problems of decision, who pursues the problem to a solution that satisfices, who recognizes satisfaction or dissatisfaction clearly, and who takes these to be criteria of his control of the situation. Another example is to be found in cost accounting, where each segment of an organization is supposed to operate within the constraints of "standard costs." When "variances" from standard occur, the part must account for them to the central administration, which then acts, if appropriate, to redesign the offending part.

Finally, many urban planners or public administrators espouse incrementalism because they believe that the designer can only influence the decision maker to change when the change is specific, not global. As the designer goes further and further ahead of the decision maker in his planning, his influence diminishes and his error increases. Hence planner and decision maker should proceed down the highway of progress hand-in-hand, so to speak.

Nonseparability

The negation of the temporary separability principle might take one of several forms, depending on what aspects of its lengthy set of assertions one wants to hold fixed. Leibniz does not seem to question the advisability of partitioning systems into parts or of attempting to control the parts. Rather he argues that the "proper performance" of a subpart depends on some concept of the whole system. In other words, he claims that the criteria of the stability of a part depends on a concept of the stability of the system S itself. Therefore, the alternative principle he has in mind is one that denies (2) and hence (3) in the list above. It asserts that one can only determine instability by examining the whole system as well as each part.

This nonseparability principle reads:

1. As before, decision rules can be constructed for the operation of each part
2'. The application of the decision rules depends on the state of the whole system
3'. There exists a function of the prior state of the part and of the whole system, which determines whether the part is stable or unstable
4. (as before) There exists a decision rule applicable to the system that will transform an unstable part into a stable one
5. (as before) As soon as a certain degree of instability of a part occurs, the designer can recognize this event and influence the

decision maker to change the system so that the part becomes stable

6'. The set of possible decision rules of each part, plus the transformation rules that send a part from an unstable to a stable state according to 1, 2', 3', 4, and 5, contains at least one member that is superior to any other rules for operating S relative to S's objectives.

I said earlier that our recent heritage leads us to be more sympathetic with the first solution to the problem of stability. But older traditions and two very modern developments would reverse the preference. We don't need to be reminded of Plato's typical attack on system design: from the general idea of good to the specific details of goodness. We have already reviewed the strong insistence of the seventeenth-century rationalists on some form of the non-separability principle. The two modern developments are cost-effectiveness analysis and the "decomposition principle." In cost-effectiveness analysis, the concept of a "part" of a system becomes quite flexible and is often called a "program." Instead of regarding an organization in terms of fixed divisions—like air force, navy, army—one thinks in terms of a classification of the major activities or "missions." Each mission requires various kinds of resources, manpower, capital, etc., which become the "costs" of the mission. A proposed mission is then compared with the other programs in terms of its whole-system effectiveness. A new proposal becomes, in effect, a proposal for the reconstitution of the system into a new set of parts. See, for example, Hitch (1965).

In the decomposition principle, each part makes its own plans in terms of the cost of resources according to its own approximations. All the plans are forwarded to one central point, which in effect compares the various demands on the whole-system resources; if a scarce resource is requested by several parts, the price of the resource goes up, and each part is informed of the new prices. On the basis of this new information, a redesign of each part is proposed, and again the new plans are compared. Under certain assumptions this procedure leads to an "optimal" overall plan. The point is that the computation required to accomplish the whole plan at the outset may be enormous, whereas the smaller part-plans may be computationally feasible. See, for example, Dantzig and Wolfe (1960).

It will be seen that both cost-effectiveness and decomposition planning satisfy Leibnizian conditions: the optimum plan of a part is a function of all the other plans, and a change in any part may require a new evaluation for all the rest. Nevertheless, neither method has been applied to systems larger than corporations or federal agencies. Part of the reason, of course, is that for larger organizations there is no central agency that can compare competing plans and work out appropriate cost and effectiveness measures. Such a center would have to translate objectives like health, education, poverty, safety, into some common social welfare units if it were to try to use either of the two methods, and it would have to have the responsibility of enforcing its overall criterion. Few seem willing to put so much power into one part of the system, chiefly because there would be no way for the other parts to exercise control over the central agency's computations.

The Reality of the Whole System

To speak of centralized control reminds one of all forms of "statism," from violent fascism to peaceful socialism. Liberal democracy has long fought any such trend toward overall system considerations. It is based on a pluralism of competing values. Many liberals have claimed that "there is no such thing" as a group mind, or a state mind, or—I suppose —a whole system. Part of this claim arises from their strong empiricist attitude that something which cannot be directly apprehended cannot "exist," and part arises because of their deep-rooted belief in the ultimate value of the individual. And yet to a philosophical "monist" they fail to show how an individual mind can exist without a group mind, or how an individual can judge himself except in terms of his society. And, indeed, the group mind does seem to exist, even in a liberal society. Further reflection on the government process just described reveals that the Bureau of the Budget is a more general designer than any specific agency; it recommends allocations of resources to Congress based on an overview of the government's activities. Congress, too, is a partial decision maker, partial designer. It implements certain programs, but also recommends others to the public via its many constituencies The public also is partially the designer, partially the decision maker. With respect to the international system, a nation is partially designer, partially decision maker. Hence, it is possible to view "designer" and "decision maker" in such a way that these do exist and do encompass something that begins to look like a "whole system."

For the rationalist, the whole system must be a reality and, indeed, the highest reality, the *ens realissimum,* that gives meaning to the existence of all things. In the beginning of his *Ethics* Spinoza considers the reality problem of the system designer: what can he assuredly assume to exist? Now the properties of any part of a system like its weight, color, etc., do not imply the existence of the system, because we can easily imagine that these properties change and no longer exist. But, argues Spinoza, substance is something that cannot be conceived in any other way than through itself, and the designer is completely incapable of designing away substance, which is Spinoza's word for the highest reality. Later history seems to have gained no insight from Spinoza as to how we can legitimately use this idea to provide the basis for the design of a system guarantor.

Leibniz is stern in his condemnation of those who want to make reality reside only in the parts, and who wish to design systems by starting with the simpler problems and working up to the more difficult:

. . . I don't want to prejudge people's intentions and therefore I don't morally criticize philosophers who wish to get rid of purpose in science and systems design. But nevertheless, I must confess that the consequences of their position seem to me to be quite dangerous. They believe that there is no such thing as an overall system or a most general system. On the contrary I hold that it is in the concept of an overall system and its performance that one will find the underlying principle of every part and the performance characteristics of the part, because the overall system is the only standard of good or excellent performance. I want especially to emphasize that one only ends in confusion when he tries to determine an optimal plan solely in terms of some particular design of a component, as though the whole system had only this part to be concerned about, instead of its entire operations.[1]

Thus the rationalists not only demanded a perfect whole system as a

[1] "Comme je n'aime pas de juger des gens en mauvaise part, je n'accuse pas nos nouveaux philosophes, qui prétendent de bannir les causes finales de la physique, mais je suis néanmoins obligé d'avouer que les suites de ce sentiment me paraissent dangereuses, surtout si je le joins à celui que j'ai réfuté au commencement de ce discours, qui semble aller à les ôter tout à fait, comme si Dieu ne se proposait acune fin ni bien en agissant, ou comme si le bien n'était pas l'objet de sa volonté. Je tiens au contraire que c'est là ou il faut chercher le principe de toutes les existences et des lois de la nature, parce que Dieu se propose toujours le meilleur et le plus parfait. Je veux bien avouer, que nous sommes sujets à nous abuser, quand nous voulons déterminer les fins ou conseils de Dieu, mais ce n'est que lorsque nous les voulons borner à quelque dessein particulier, croyant qu'il n'a eu en vue qu'une seule chose, au lieu qu'il a ne même temps égard à tout." [Leibniz, *Discours de Metaphysique,* paragraph 19]

standard for system design; they also demanded a fully qualified proof of the existence of a perfect whole system.

We note, then, that the rationalist has added one very important property of the "whole system" that was omitted as we described the expanding decision maker-designer above, namely that the whole system is basically good. That is, we could not say that collective mankind is the real whole system unless we could show that its intentions were good. If we could do so, we might then speculate on whether it is appropriate to call such a collective "God."

The rationalist was in his own view a scientist, and for him nothing can be admitted to the fund of knowledge that has not passed the most carefully designed criteria of objective truth. But for him "objectivity" rests ultimately in the concept of a benign governance of systems. Today, theology and science have no common meeting ground. A theology that must postulate a God rather than provide a precise proof of His existence cannot expect to find acceptance as a branch of science in an age when the essential feature of science is its strict adherence to standards of precision.

Now the rationalist did believe he could find a precise and simple proof of God's existence. This does not mean that it is a simple matter to find such a proof, as the tortuous passages in Descartes and Leibniz clearly show. But all who have tried mathematics have had that quite wonderful experience of finding, after hours or years of labor, a very simple way of proving something that was not obvious at the outset.

The failure of the rationalism of the seventeenth century lay in its inability to find a simple and precise proof that all could agree to. It was Kant who finally exposed the fallacies of all proposed simple proofs. The essence of the Kantian refutation was that the conceptual framework required by science to give meaning to experience was not logically strong enough to establish a God in the sense demanded by a Leibnizian theory of reality. In the post-Kantian period, Hegel attempted to revise Kant's notion of this conceptual framework, and thereby to establish an Absolute Mind. His Absolute Mind plays exactly the role required of a "whole system," because it establishes the grounds for meaning in any aspect of reality. But western science, at least, could not tolerate the ambiguities of Hegelian logic, which required contradiction as a necessary condition for proof. Today contradiction still plays the same role as it always has in western science: it is that which establishes the stopping point of formal inquiry.

Monism

Thus we can see the lines of the intellectual battle. It is a fight between pluralism and monism, between those who wish to see and design their world in pieces and those who wish to see and design it as a whole. The pluralist is a problem solver, incrementalist, individualist, empiricist. He becomes most uncomfortable when challenged to explain what the system is supposed to accomplish, what are its "real" objectives, because in his heart he doesn't believe that a system has objectives; "only people have objectives," he says.

Pluralism is very popular today, but to the monist it is essentially irrational. On the one hand it praises "freedom" and "individualism," and through political or even military forces tries to bring about a freer world. Hence it does believe in overall objectives. On the other hand, it won't defend its policies, even when in the attempt to establish freedom so many people die and lose their freedom.

Since pluralism has so much the flavor of common sense, it will be worthwhile to spell out in more detail the meaning of monism as the rationalist sees it. The overriding idea is that existence has a purpose, and that the purpose is good. There is, in fact, a guarantee that, despite heartache and body ache, we are on the road to blessedness. This is the best of all possible worlds, either because it is the only possible world (Spinoza) or because it was designed by a perfect designer (Leibniz).

Monism says a great deal, and our modern-day attitude toward it may be to reject some of its tenets and accept others. The monist's tenets are all concerned with the concept of the "whole system." A "whole system" in its broadest sense is that system of which every other system is a part. This implies that the goal of the whole system sets the goals of every other system, since according to the definition of a system given above, "part performance" is always evaluated in terms of "system performance." The pluralist, of course, does not believe that such a whole system exists, i.e., he does not believe that there is any system that sets the performance standards of all systems.

The basic tenets of monism are (1) the whole system exists and is unique; (2) the whole system is optimal; and (3) the proof of the existence of the whole system and its properties meets the requirements of scientific proof (i.e., has the highest level of objectivity).

Proposition 1 contradicts the pluralistic philosophy of conventionalism. Few pluralists would question the *convenience* of using constructs that enable us to integrate our empirical findings. For example, students of organization theory often act as though there really were a total organization "out there," just as political scientists sometimes seem to act as though there is such a thing as the federal government, and engineers as though a total generator plant existed. To a strict empiricist, however, these suppositions are merely convenient ways of tying together a series of observations. He would not permit the scientist to claim reality for his construct, since the construct is never observed. The monist, of course, goes far beyond the organization theorist in his claim that the most comprehensive system exists.

The empiricist, as we shall see, adopts one answer to the problem of the ontological status of sense impressions, an answer that he takes to be based on the principle of parsimony in inquiry: never accept any more than is strictly warranted by the sensory evidence. To the monist the difficulty with his answer is demonstrated in the discussion of rationalism. If the empiricist is telling us how to design inquiring systems, then we must ask whether the kind of parsimony he requires is desirable. The answer to this question depends on the manner in which the inquiring system gathers its evidence, i.e., on the design of the whole inquiring system. In other words, if the scientist makes choices, e.g., between parsimony and richness, what dictates the choice he makes other than considerations of the objectives of the whole system?

Proposition 2 is not interpreted in the same manner by all monists. In Leibniz there is a perfect system, God, which designs the whole system; the remaining parts of the whole system are imperfect, and their "measure of performance" is gauged in terms of God's perfection, i.e., God's teleology. The optimal whole system, therefore, contains less than perfect parts, a mysterious necessity ingrained in reality. In Spinoza, the whole system is perfect and there is no designer. As is well known, Spinoza seems to contradict himself in the *Ethics* by implying that human beings "ought" to increase their understanding of the whole system, for this prescription leads one to conclude that the human mind is not perfect. If everything that happens takes place because of the perfection of the whole system, then this must be true of understanding: we understand just what the nature of the whole system implies we are to understand. Spinoza's problem is a subtle one, but a further discussion of it would take us too far from our interest, because Spinoza is not concerned with the design process, while Leibniz clearly is.

As was indicated in the last chapter, one role of the perfect entity is to guarantee the survival of the whole system. In philosophical tradition, X is perfect if it is not limited in some respect. In other words, the general property "good" can be subdivided into a set of properties: intelligent, beautiful, knowing, powerful, and so on. Entities having these properties can be ranked so that, for example, "is more powerful than" orders the objects of the word. For each such property there is a maximal entity, e.g., an entity which is more powerful than any other entity. Finally, it is asserted that there is but one entity that is maximum. A most intelligent entity is also most powerful and most beautiful. The *ens realissimum* is that entity.[2] In Leibniz, proposition 2 is true by virtue of the fact that the perfect entity exists.

The pluralist wishes to restrict the whole system to what seems to be practically conceivable, and would of course deny this assertion. He would, in effect, argue that the theory of systems design does not have to commit itself concerning the properties of the most general system, and certainly does not commit itself to the existence of a perfect system. In particular, the pluralist would not agree with the uniqueness of the maximal entity along all value scales. Most systems designers go as far as they can in trying to conceive a system that will be best for some specific purpose. But they do not feel that these systems are best for all purposes. A missile system may be designed by the designer trying to conceptualize what an ideal missile should do, e.g., it is one that destroys an enemy stronghold perfectly. If so, he does imagine a "perfect" system within the limits of his imagination. But he would hardly say that the missile system was perfect in all respects. It is not very effective for producing consumer goods, for example.

The Leibnizian answer is fairly obvious, of course. It simply says that for every missile designer there must be another systems designer who considers the missile system as a part of his system. Such a designer also tries to conceptualize the perfect system. For him, the ideal missile may not be one that destroys perfectly. It may, instead, be one that prevents destruction perfectly. In this case, the original missile designer made a mistake in his selection of the relations that rank entities. In short, only if one conceptualizes the most general system will one know what relations are appropriate in ranking entities.

---

[2] The axioms are as follows: Let [x] be the set of all entities of the world and [R] be a set of relations: (1) Every member of R is asymmetric, transitive, and closed with respect to [x]; (2) for every R there exists a (unique) member x_{max} of [x] such that $x_{max}Ry$ is true for all y in [x]; for all R_1 and R_2 in [R], the x_{max} of R_1 is identical to the x_{max} of R_2.

The unique and perfect system concept can be understood in another way. In our culture, we typically segregate the functions that men perform, in terms, say, of the professions of research, law, education, industry, government, and so on. The professions come in contact only on the periphery, so to speak, where a man of one profession consults a man of another. In the consultation, the one learns about the results of the other's deliberations, but does not take a hand in framing the results. This is essentially a partitioning of our social institution into presumably separable parts. Each profession can be understood by itself, by understanding the manner in which it works and the principles that guide its actions.

But suppose one were to deny all this segregation of the professions and were to say, for example, that one cannot understand science unless one has understood it as a management profession, a political activity, or a legal activity. For example, one might argue that science can manage an enterprise, or a part of it, and that operations research is just such a way of viewing science. One might further argue that there is some optimal way in which science can manage: an ideal of scientific management. Finally, one might argue that a necessary condition for understanding what science is, is the understanding of how it can and ought to manage.

It must be emphasized that all along we are discussing the design of systems. Hence, the question is not to understand how present-day science can manage, because present-day science is a very imperfect system. The question is: what would science have to be like in order for it to be a management? Once one begins to understand this question, he is on his way to understanding God, the perfect system. For God can be looked at as a perfect scientist; but in the very perfection of God as a scientist lies His perfection as a manager.

By the same token, a necessary condition for a full understanding of management is to conceive of management as a science. This, indeed, is what is happening in the current developments in research and development, where management is playing a stronger and stronger role in the planning phases of research. There is an activity called the management of science. There is an optimal way in which management becomes a science, i.e., a generator of information. To understand management, one must understand it as a science. In other words, one way to understand management development is to determine in what way management can become a scientific system. The way in which God is a perfect manager can be understood by the way in which God is a perfect scientist.

The same theme could be repeated in many contexts. To understand science, one must understand it as a legal profession, and to understand the law, one must understand it as a science. For example, T. A. Cowan (1948) argues that law is the system of controls for experimentation in the social sciences. It seems to me that he is trying to conceive of law as a science. I know of no one who has yet been bold enough to suggest how science becomes the law, except in the bad sense of a science that controls thought processes. But such an attempt needs to be made if one is to understand God as the perfect scientist: the manner in which He is a perfect scientist includes the manner in which He is a perfect lawmaker.

Even within the scientific disciplines themselves, the same principle could be applied. For example, one cannot understand psychology until one has understood in what way psychology is a physical science, i.e., has understood how a perfect psychology must include a perfect physics. One notion of how this might come about is to "reduce" all mental phenomena to neurological phenomena. But such reductionism is not the whole story; we must also show in what sense physics is a psychological science.

We see in what manner present-day science is so imperfect. In Leibniz's terminology, the scientist's apperception is very weak, even though his perceptions may be strong. Interpret Leibniz's "apperception" to mean the ability of the system designer to design the system from many points of view—to design science as a management system or design physics as a psychology. To the extent that a system fails in its apperception, it is less than God, i.e., it is an imperfect monad.

Thus the implication of the rationalist thesis is that the system's designer does not understand his system until he understands it in terms of all the basic functions. The designer of a missile system must understand how the missile system is a productive system, a communications system, an inquiring system, and so on, *if* he is to understand his system fully.

The implication of proposition 2, therefore, is that: (1) all systems have the same set of functions; (2) their objective is to increase the clarity of their apperception; (3) a perfect system does exist in the sense that there is a system with perfect apperception.

Clearly, proposition 2 demands some reasonable taxonomy of systems in order to be usable. If systems of type x must be understood as systems of type y, what ranges of concepts do x and y entail? Our present taxonomy, which has grown out of the tendency to separate functions, may be far too awkward to apply.

Proposition 3 states that the existence of the whole system and the perfect system can be proven objectively. Clearly the monist does *not* mean that "existence" is to be defined in terms of "observed" or "observable." Instead, for something to be taken to exist, it must be assumed essential in the development of inquiry. He asserts that one cannot separate out segments of inquiry and stamp "existence" or "reality" on these alone, because these segments exist as segments only by virtue of the whole system. We never know what really exists, but at any time we do the best we can to construct an image of the world in which our observations, our thinking, our feeling, our intuition will live as well together as possible. We *take* such an image to exist; but it is so taken only because we argue that there is a whole image of which ours is an approximation.

Propositions 1 and 2 state a hypothesis about reality: namely, that there is an *ens realissimum*. We cannot agree with the rationalist that the proof of the hypothesis is simple. Indeed, it never will be proved because it is the most complicated hypothesis possible.

The Science of Theology

The effort to satisfy proposition 3 falls within theology. We have seen that the type of evidence required by a science of theology is quite different from the evidence of the so-called empirical sciences; the evidence for a proposition x in theological science is of the form "x is needed in order to guarantee y" where y is some evidence in a "contingent science." Although the rationalists agreed that a theological science is fundamental, in the sense that no science can exist without it, as we have seen they disagreed as to the nature of the fundamental entity of a theological science, the perfect Being.

The underlying theory of the rationalist theology was that systems can be described in terms of their "levels." For example, two individuals in a division of a company are arguing about whether the company should advertise in a cheap "man's magazine." The argument is to be resolved by going to that higher level of the corporation in which benefit vs. dignity can be compared. Or two states of the United States disagree about the use of water in a river that flows between them; a "higher" level of federal authority may step in to resolve the issue. In international crises, as well as in many disputes at the state and local levels, the appropriate "higher level" is not easily found, and, indeed, is a task of system design. Clearly the relation "is at a higher level than" is fundamental in theological science, and plays a role in all sciences that somehow try to distinguish between levels of life.

Spinoza's concept of level relies on the logical concept of inclusion. In extensional terms, S_1 is at a higher level than S_2 if every entity in S_2 is included in S_1 but not conversely. In intensional terms, S_1 is at a higher level than S_2 if every property of S_2 can be defined in terms of the properties of S_1 but not vice versa. The highest-level system in Spinoza is God, i.e., substance, which means that every lower-level system is "in God" both extensionally and intensionally. Spinoza's theory of levels more or less accords with some modern-day concepts of organizations, especially in "planned societies" where every issue between two systems at the same level can in principle be resolved by a higher level which includes both subsystems and defines their functions.

A different approach in organization theory is to consider the higher level to be distinct from the lower, and to have certain properties that qualify it to make the higher-level judgments. Ideally, the highest level must have the ability to discern all possible pathways, the objective basis of resolving all value differences, and the power to enforce its judgments. It is omniscient, optimal, and omnipotent. The higher levels need not be like the lower ones, either in their structure or their mode of operation.

Leibniz introduced a synthesis of these two opposing points of view. In Chapter 1 we considered the question whether every system has parts that are also systems; if this is the case, then there is no ultimately "simple" system. Leibniz's system theory posits the existence of simple systems that have no parts; nevertheless, we have noted that all Leibnizian systems manifest the same functions, though in all but God the functioning is imperfect. Thus all simple systems are alike with respect to their underlying types of change.

The lesson of Leibniz's monadology for system design is that all systems can be analyzed into sets of irreducible subsystems, that every such irreducible subsystem has the same set of functions as every other, and that the subsystems differ one from another in the effectiveness of their functions. Leibniz does not provide enough about the internal workings of his monads to carry the analogy into modern system design, but his point has been made many times in discussions of "general systems theory," in which students of systems try to describe the essential characteristics of all systems. See Boulding (1956).

Leibniz's idea seems to have been first proposed in western thought by Anaxagoras, but with a radically different intent. In Anaxagoras, reality is made up of a rich variety of properties, and in every real entity all these properties occur (with the exception of the mental property). Many writers, including Leibniz himself,[3] would probably assign the philosophical origin of the principle to Plato. Certainly Leibniz's theory of ideas is Platonic; and certainly Anaxagoras' theory is materialistic. And yet it seems to me that the credit for the central idea—that one cannot describe any piece of the real world without using all the basic descriptors of the whole world—must be given to Anaxagoras.

What remains to complete the Leibnizian theory of system design is a classification of the basic functions of all systems. There also remains the problem of whether or not there are ultimately simple systems. Finally, there is the overriding problem of the meaning and existence of a system guarantor. The last stipulation on page 43 remains a problem, and raises the question of whether systems "exist" in a pragmatic sense: How can the designer ever be satisfied that his efforts amount to anything that is really worthwhile?

The problems we have gathered from the rationalists will be considered in further depth after some more historical excursions.

[3] See "Briefe an Vic. Remond" in Leibniz (1914).

: 4 :

THE LEIBNIZIAN INQUIRER
ILLUSTRATED:
ORGANIC CHEMISTRY[1]

Thus far we have said little about what many philosophers would regard to be the central problem of inquiry, namely, the justification of induction. The problem is usually formulated as justifying the inference to a generalization from the observation of specific events. In Chapter 2 we discussed a Leibnizian inquiring system in which the confidence in any assertion, general or specific, increases as the assertion becomes an integral part of a large fact net, i.e., tends to lie near the bottom of the net. If this rather vague stipulation could be made more precise, then one solution of the problem of induction might emerge.

In the last chapter it was indicated that observation is probably not the "starting point" of inquiry because it is not a separable component of the inquiring system. Even if this is true, the problem of induction still remains, namely, to indicate how the inquiring system can relate specific "facts," no matter how contingent, to generalizations.

In this chapter we consider a specific example of an attempt to design a Leibnizian inquirer, in which the inductive problem is central. The example lies in the area of analytic organic chemistry. This selection was based in part on the principle of the mean between extremes: we wished to avoid an oversimplified example that is of no real concern to anyone (e.g., are all swans white?) and, on the other hand, an example so complex that no one really understands how the system works.

Before describing the components of the system we tried to design,

---

[1] The material in this chapter is based on a research project conducted by E. A. Feigenbaum, Joshua Lederberg, Bruce Buchanan, and others, together with the author, at Stanford University. An earlier version of the chapter, entitled "On the Design of Inductive Systems: Some Philosophical Problems," was written by Buchanan and the author (1969).

suppose we begin by representing the so-called "inductive process" as a two-component system. The first component roughly corresponds to the discovery process, the second to the confirmation process:

1. Find an H which satisfies the schema, D because H and E, given D, the data to be explained, and E, the background statements, and some specified sense of "because"; and

2. determine the degree to which hypothesis H proposed in 1 is satisfactory, given the data and background statements, for some specified sense of "degree of satisfactoriness."

In recent philosophical literature, different kinds of hypotheses may count as relevant answers to why-questions. For example, a "how-possibly" explanation (in Dray's sense) may be appropriate in some contexts while other contexts call for deductive-nomological explanations (in the Hempel-Oppenheim sense). That is, different senses of "because" may be applicable in different contexts, i.e., the links of the fact net may vary. And, perhaps more obviously, the same is true for satisfactoriness. A hypothesis which is intuitively very likely may have a low degree of confirmation in some logic of confirmation.

In the language of the last chapter, "because" and "satisfactory" are related to the components' measures of performance, which in turn describe the goals the system is supposed to attain. In this simple, two-component inductive system, the closer the first component comes to meeting the requirements of a specified "because" and the higher the degree of satisfactoriness (in one specified sense), the higher the measure of performance of the inductive system.

Hence the designers of methods for formulating and testing hypotheses, i.e., of inductive systems, must not only specify methods for carrying out 1 and 2, but they must specify what they mean by "because" and "degree of satisfactoriness." By specifying what he means by "because" in his system, a designer is determining the kinds of hypotheses which he is willing to admit as plausible or relevant candidates. (Cf. Hanson, 1961.) And when he specifies criteria for determining how satisfactory a hypothesis is in his system, he is determining the degree to which hypotheses will count as satisfactory answers to the why-questions posed to the system. In fact, much effort in the philosophy of science has been directed toward explicating the terms "explanation" and "degree of confirmation." But relatively little attention has been paid to the total inductive system for which these explications are made.

Now there are a number of ways of conceiving the problem of designing an inductive system. One very prevalent way in philosophical literature is to assert that the problem from the philosopher's point of view must be framed solely within a formal language (i.e., a "logic"). In this regard, the problem is similar to the deductive task of codifying the rules which define an "acceptable" string of symbols stretching from a given set to some desired conclusion. For example, Carnap's studies of induction center on designs for a logic of confirmation. No one who holds this view of deduction or induction denies that there are fascinating extralogical questions of elegance of proof, intuitive insight, and so on. However, others, like Goodman (1965), seem to have serious doubts whether the inductive problem can be adequately encompassed within a purely formal language built out of logical connectors and terms.

As will be seen, the practical problem of induction which is described in this chapter leads us to a very definite position on this point. It will appear clear that the purely logical aspects of the problem cannot be distilled from other considerations, and especially not from the economies of time and strength of insight of the qualified scientists. In other words, quite apart from Goodman's question about the extralogical character of projectible predicates, there are essential extralogical considerations that enter into the choices the designer of an inductive system must make. We shall also see how the fact nets constructed to improve our insight into chemical processes never lead to completely satisfactory "solutions."

In mass spectrometry, a sample is fragmented by a bombardment of electrons in a mass spectrometer, thereby producing ions of different masses. One resulting set of data is a graph, the x axis of which represents masses (molecular weights[2]) and the y axis the so-called "intensities" which, roughly put, are the relative frequencies with which a fragment of a given weight occurs. Thus the graph provides us with one set of contingent truths for the fact net.

In terms of the simple scheme given above, the specific problem we consider is to explain the data, D, produced by a mass spectrometer, given E, the existing theory of mass spectrometry plus background conditions. That is: (1) find a hypothesis H (in this case a molecular structure) which satisfies the scheme, D because H and E; and (2) determine that H is most satisfactory (i.e., that no other molecules account for the data—or at least not as well as this one—under this theory and these conditions).

[2] More accurately, the abscissa points of a mass spectrum represent the mass to charge ratio (m/e) where most, but not all, fragments are singly charged.

The set E consists of sentences about the fragment, about organic chemistry, and about the conditions of the experiment supplied by chemists and others. E is a Leibnizian fact net, or set of fact nets, linked together in various ways. It contains both specific and general statements. It is created by asking chemists and other experts a series of questions, e.g., "Why do you say that a fragment of such-and-such molecular weight must have such-and-such a formula?"

The hypothesis H is not just a chemical formula, but a specific molecular structure which shows the atoms and their bonds. Even for a relatively simple formula there may be thousands of "isomers," i.e., corresponding structures. Most of these, of course, may be very unstable, and therefore could not be satisfactory candidates for H. Thus many of the sentences in E concern themselves with the theory of chemical instability.

As in other routine scientific tasks, the set of concepts to be used in the explanations and in the descriptions of the experimental results is fixed. The higher-level problem of finding a language in which to describe this "raw data" and explain them does not apply here, at least initially, though we shall address ourselves to it in the speculative conclusion of the chapter. That is, for the present we regard the inquiring system strictly from the point of view of chemistry; we postpone the Leibnizian question posed at the end of the last chapter, namely: Can this system be regarded as a non-chemistry system?

The fact net was designed into a computer program. In the system embodied in the program the H's which satisfy the schema in (1) are molecular structures which are consistent with the data to be explained (where a simplified theory provides a consistency check). The theory of mass spectrometry currently is not sufficient to allow one to incorporate the deductibility criterion of deductive-nomological explanations. That is, the link between the "facts" in E and the contingent truth H to the facts of the mass spectrograph is not a deductive link at this stage. Instead, the most "satisfactory" hypothesis H in this system is one with the highest estimated "degree of satisfactoriness." The program makes this estimate by scoring the seriousness and the number of mismatches between a prediction for a candidate molecule and the original data— using admittedly incomplete methods of calculation.

We may note at this point that H is in the form "X has property P," where X refers to a particular, individuated object, namely, the sample the chemist used in the mass spectrometer, and the property attributed to X is a graph structure.[3] Seemingly, therefore, our problem is quite different from the "classical" problem of induction, which was concerned with the method of passing from data to universal propositions. We say "seemingly" because, as we shall see, it is quite impossible to separate one's attitude toward the correctness of H from one's attitude toward some of the tentatively offered universal sentences in E, the set of background statements provided by expert chemists. A very successful induction to a specific H increases the confidence in the theoretical laws. This result is very much in accord with E. A. Singer's description of the empirical process of science (1959), and W. V. O. Quine's metaphorical description of the interconfirmations between laws and particular statements (1953); a geodetic surveyor's success in measuring the distance between two specific points on the surface of the earth at a specific time increases his confidence in the theory of the instruments he uses.

Chemists who work on the problem of identifying molecular structures from mass spectra have no rigorous or mechanical procedures which lead them from data to hypothesis. They look at clues, make guesses, and reject hypotheses in typically undefined ways. Thus there seems to be no explicit guide to the design of a satisfactory fact net. Our task was to determine whether these many intuitive moves of the chemist can be designed in an explicit way.

In order to understand the system, it will be necessary to enlarge the number of components, i.e., the number of distinct system tasks. Each component, except the first, has been designed into a computer program. The tasks of these components, in general terms, are:

1. Collect data
2. Adjust data in light of current theory
3. Suggest classes which contain plausible hypotheses by looking at the most significant features of the data
4. Suggest further limitations on hypotheses after consulting outside experts or the results of other tests (increase the theoretical base)
5. Construct plausible hypotheses (using 1 to 4)
6. Make predictions for each candidate hypothesis
7. Assign degrees of satisfactoriness to the candidate (using 6)
8. Recycle if no hypothesis is "satisfactory enough."

---

[3] Thus the problem to be solved can be regarded as a mapping of one set of graphs (the mass spectrographs) onto another (the graphs of molecular structures). We return to this representation of the problem at the end of the chapter.

It is by no means obvious that this set of items constitutes components of the inquiring system in the sense of the last chapter; e.g., it is not clear that we can construct measures of performance of each component such that an increase in the component measure produces increases in the whole system performance. But again for present purposes it will be wise to postpone some of the systematic problems of our inquirer and to act as though this breakdown were satisfactory.

Step 1, data collecting, is not an easy step to mechanize fully. In a complete design, it is essential to specify not only how to collect data but what data to collect. Researchers have documented many problems encountered in this stage and have proposed various techniques for solving them, for example, random sampling techniques. However, it still is not clear what criteria a researcher should use to decide, for example, when one instrument or technique will give him data better suited to his immediate purposes than others, when to stop collecting data, when the conditions are unfavorable for collecting reliable data, or even what will count as a datum.

It would appear, however, that step 1 seems to create a system boundary, where data are "received" from outside, and to this extent the inquiring system is not Leibnizian. At this stage of development, the comment is valid; the operation of the mass spectrometer and the selection of a sample are, in fact, in the environment of the system and not part of the design. But ultimately these aspects would have to be swept in. Certainly the origin of the sample is important to the chemist, as is the underlying theory of the mass spectrometer. In the end we are led to speculate on what is truly "given" to the inquiring system.

Even the present system is not altogether passive with regard to the data. In component 2 it sorts out the uninformative and unreliable data and finds the most significant. Certain of the spectral lines on the graph have no significance, either because they are produced by the machine and not by the sample or because they are produced by every sample. Thus the designers must specify for the machine exactly what they mean by "uninformative," "unreliable," and "significant." Although chemists have been able to give explicit criteria satisfying some of their intuitions, there seems to be a real possibility that other criteria will improve the performance of the whole system. Thus the current theory of mass spectrometry indicates that some of the data (peaks in the bar graph) should be ignored because, if present, they must be due to impurities in the sample, lack of care in recording data, a faulty instrument, or something of the sort.

We should now note that this component of the system is not separable from other components, and especially not from the fifth. If we tried to define the second component's task in purely logical terms, we might try to define "degree of relevance" of a datum, and design the component so that it accepts all data with a relevance greater than a virtual zero. This might make the fifth task of constructing plausible hypotheses much easier, but on the other hand it would result in making the tasks of the first four components extremely complicated. There seem to be no purely logical criteria for deciding which data to use at step 2, but there may be some economic criteria based on cost and time, as well as technological criteria based on computer memory capacity. In other words, the designer is forced to employ a mixture of logical and extralogical considerations in *selecting* the data to be used. This seems a far cry from the affluent society where a man uses the observations of every swan he sees to check the hypothesis that all swans are white. But even where such affluence exists, it may be a poor strategy to restrict oneself to the language of the hypothesis in making inductions.

In order to explain this last remark, it will be helpful to look ahead a bit and describe E. A. Singer's paradigm of scientific method, which we shall examine in more detail in a later chapter. We suppose that the inquirer has a capability of assigning one of a set of properties to identifiable objects by means of observation. As in the illustration we have been considering, the inquirer wishes to select a satisfactory hypothesis about these objects, e.g., that all objects of a kind have such-and-such a property. Let us suppose that the inquirer is remarkably successful, so that indeed every object he examines has predicate P without exception. Would such a situation provide a satisfactory hypothesis? Most philosophers, including Singer (1959) and Goodman (1965), would say no, because what is needed besides is some "explanation" of the success; Goodman would wish to test the projectibility of the predicate, and Singer would wish to subject the hypothesis to a more critical test. The point that each makes is that mere repetition of "successful" observations does not provide any (or provides very little) additional information. This is so because there may be a large number of explanations of the hypothesis, none of which is excluded by the dreary repetition of successful observations.

This situation is dramatically illustrated in our study. Suppose the inquirer were to concentrate only on a small set of mass points (x points on the bar graph), and were to ignore all the heights of the lines (y points). Thus the second component of the system has performed its task in a very economical way. The third component now seeks to find classes that contain plausible hypotheses that will account for the existence of these few mass points, and the fifth component seeks to find in these classes those specific molecular structures which, if fragmented in the mass spectrometer, would produce at least fragments of these masses. When a candidate is found, it is confirmed that it satisfies the conditions of the problem. This means that every time a sample with this candidate structure is fragmented, these selected points appear on the x axis of the resulting bar graph. But the trouble is that there are thousands of other candidates which also have this property. Selecting an economical set of data reduces our ability to discriminate among alternative explanations. Moreover, since the same selection criteria are used in each experiment, a mere repetition of experiments (on samples drawn from the same population) does not increase our ability to discriminate.

Singer's proposal is that the inquirer partition the predicate P into a set of *exclusive* predicates $P_1, P_2, \ldots, P_n$, such that for $n > 2$, every object having predicate P_1 has predicate P, no object has both predicates P_i and P_j $(i \neq j)$, and every object having predicate P must have one of the predicates P_1 in the partitioned class. (The most familiar way to accomplish partitioning is to let P be a scalar quantity, and the P_1 represent the next significant decimal place; but partitioning can also be accomplished when P is nonscalar.) Now the explanation of chief interest to the inquirer must be such that it enables him to say which predicate of the partition class the object must have if the explanation is true. Thus successful partitioning may enable the inquirer to discriminate his candidate from other candidates which predict other predicates in the partitioned class. We say "may," because mere partitioning by itself does not guarantee such discrimination; there must be some theoretical assertions which provide this guarantee. Furthermore, the ability to observe the predicates of the partition demands refinement of instruments and increase of the theoretical base. Evidently the partitioning step can be repeated at the next level, and the grand hypothesis is that at some level of partitioning one and only one "satisfactory" explanation will emerge.

Thus Singer puts some life into the Leibnizian fact nets. Instead of merely processing a set of "facts" in a passive manner, the inquiring system sets up its own requirements for data. The battle between competing fact nets becomes quite impressive. We note in our example that if there are two strong candidates for the "satisfactory" molecular structure, then each will be attached to a fairly elaborate fact net, containing many assertions about stability, about the manner in which fragmentation occurs, and so on. Indeed, looking ahead, the process is quite Kantian in flavor. At the lowest level (we call it the "first-order theory") some elementary a priori theory enables the inquirer to attach predicates to objects and sift out classes of plausible hypotheses. As we proceed to partition the predicates the theoretical a priori base must be increased. In our case, there are several ways we can partition. We could use mass spectrometers with a higher resolution, which would enable us more accurately to pinpoint the composition of the molecule and fragments, or we could bring in more of the original data (y coordinates of the bar graph as well as more x points). In fact, both were done, but for this discussion it will suffice to describe the latter. At the first level of partitioning, there was a fairly crude method of classifying the intensity lines in terms of "high," "low," and "absent." The expert advice of our fourth component, which suggests further limitations on the hypotheses, must then pour into the system various sentences which provide the implications of high or low intensity lines. We were, in fact, probing the more or less unwritten lore of mass spectrometer chemistry. Some of the assertions given us by the chemists were by no means obvious to the layman; for example, that if spectral lines exist for fragments of mass x_1 and x_2, then the original sample was an instance of a structural class (e.g., ketones). In other words, once we had introduced a refinement in the data, we were able to ask more penetrating questions of our informants, and hence, as it turned out, to discriminate far better among the candidate hypotheses. We then proceeded to partition even further by refining our classification of the heights (y points), and thereby adding more background statements supplied by chemists.

To the purist, the fourth component must seem very sloppy and haphazard. This component identifies certain people as experts, and then asks them to supply rules to the inquiring system which help the system to discriminate better, so that in the fifth component the number of specific plausible hypotheses is significantly reduced. This aspect of our design problem does distinguish our efforts from those which are concerned with making inductions de novo. It would be fruitless to try to design an inductive system which learned all its organic chemistry and mass spectrometry only from its own observations. No one expects the chemistry student to learn in this fashion. The problem is rather of the following type: Given a fairly rich set of theoretical sentences which have a relatively high degree of implicit or explicit acceptance within a discipline, how does the inquirer become aware of these sentences, and how does he select the appropriate ones to increase his power to partition and discriminate among candidate hypotheses?

Of course, the expert may also have nontheoretical information about the sample in question which the system can use in this specific case. For example, he may know that the sample is probably an aldehyde derivative because it was synthesized from an aldehyde. Often he brings in results of other analytical tests—infrared spectrometry, for example. Or he may know what sorts of things the synthesizing chemists were interested in, or he may read clues from the smell of the sample, its boiling point, or other features.

Again it should be noted, the components so interact that clear a priori guides to system design are lacking. For example, it would be possible to avoid steps 3 and 4 if the designers were willing to invest time and effort in steps 6 to 8. That is, the program could consider all the possible hypotheses as plausible, counting on disconfirming the unacceptable in the prediction-comparison phase. Or, conversely, by putting much more effort into step 3, it might be possible to eliminate the prediction phase by reading clues from the data so closely and carefully that there remains only one hypothesis in the class of plausible candidates. It is not clear how much effort the system should put into one phase or the other—and not even clear to the designers how to decide the issue. Intuitive notions of efficiency had to be the guide for the most part.

We should note that generally speaking, all the contributions to the inquirer from components 1 and 4 are "contingent truths," because the inquirer does not have to accept any sentence as final. Of course, a sentence acceptable to all chemists belongs way down in the fact net of organic chemistry, and therefore has a very small chance of being rejected because of its invalidity, though it may be rejected for its irrelevance.

If we return to the problem formulation given at the beginning of this chapter, we can reformulate the task of the first five components of the inquirer as follows:

Find as small a number of H's as possible which satisfy the schema, D' because H and E', where D' is some partitioned subset of the original data base and E' is the union of the original E (theory and experimental conditions) with the statements supplied by experts.

At this point the system must be able to generate hypotheses which are consistent with the (significant) data, according to the theory, and which satisfy other constraints. In the case of this program, the total hypothesis space is defined by Lederberg's DENDRAL algorithm (1964, 1965), which generates all acyclic connected graphs given the number of nodes (the composition)[4] and the number of links from each node (the valences). When this algorithm is constrained by a model of chemical stability, it generates only hypotheses which are stable chemical molecules having the specified composition (isomers). Further constraints from steps 2 to 4 allow only generation of those isomers which fit the significant data and which contain, or exclude, certain structural fragments. The existence of the DENDRAL algorithm is crucial for this system, even though it always operates with constraints. Even when the system has such a hypothesis-generating algorithm, the discovery process is complex, as we have seen. In other inductive tasks where there is no known algorithm, the inquiring system would tend to be even more complex, and perhaps not designable.

We turn now to a consideration of components 6 and 7, which test each candidate hypothesis, and hence perform the second task of finding the most satisfactory hypothesis. At first glance, the task seems simple. Apparently, all one need do is to predict the graph that would be produced if the candidate molecule were to be fragmented in the mass spectrometer, and then compare, line by line, the actual graph with the predicted graph. But to perform this task perfectly would require far more theory than is now available, or likely to become available. Indeed, were such a strong theory of the mass spectrometer in existence, its impact on the other components would be considerable. Note, for example, that the third component asks whether a certain class of molecules could have produced a specific spectral line. A perfect theory would immediately tell us whether this is so, and, indeed, might list all molecules that could produce a line of a specified intensity.

Since there is no such perfect theory and hence the sixth and seventh components must operate imperfectly, their design faces much the same sort of problem as do the other components: a delicate balance of relevant data and expert theory vs. overloading the system. Indeed, the earlier tasks are all reintroduced at this stage. One must reconsider which spectral lines really are important. If the candidate theoretically produces a line at x, when no line shows on the actual graph, or fails to produce a line when a line exists on the graph, does this matter? The

---

[4] The current program takes the composition as a given piece of data, e.g., it starts with a "given" composition, $C_2H_{10}O_{12}$, say. Some work has been done on a program that would infer the composition, or a set of possible compositions, from the original data.

work of the second component is thus reexamined, as are the contributions of the experts in the fourth component.

The seventh component's task is classical: how to assess the plausibility of each candidate. It is exactly the problem faced by the statistician, who uses likelihood ratios to test the comparison of theoretical data points with "actual" data points. We merely note (again) that the classical version of this problem does not fit our needs because of the ever-present problem of the second component: which *are* the "actual" data points? The statistician needs to have some probability density associated with each spectral line, and this seems difficult to provide at the present stage of theoretical development. It may eventually be possible to use a Bayesian approach, but in this study a much simpler scoring system was chosen, which operates on the spectral lines selected by the second component. Thus each candidate receives a score, which in turn either rejects the candidate or permits a battle to occur between candidates and their associated fact nets.

The last component, which recycles if no "satisfactory" hypothesis is forthcoming or if a battle develops, was not programmed on the computer, partly because there seemed to be no clear-cut strategy for its operation. What is needed is some history of the weakest links in the total process which might account for failure, so that in the recycling the inquirer could systematically change the weak steps to see whether a satisfactory candidate emerges or which candidate seems better. Including the chemist as part of the system alleviates this difficulty, for he is able to draw on much more experience than the program, and thus has a "feel" for when something has gone wrong.

The example of an inquiring system we have been examining provides a rich background for speculation. Of chief interest for this book is whether the program is a system. Suppose we return to the necessary conditions described in the last chapter.

1. The program is certainly teleological, in the sense that it makes choices to arrive at a specific goal—the most "satisfactory" candidate molecule.

2. On the other hand, we did not succeed in arriving at a suitable measure of performance. Note that even had the scoring system ("degree of confidence") of the candidate molecules turned out to be excellent, it would still not be a suitable measure of performance for the whole system. The scores at best tell us whether the system has generated one and only one satisfactory molecule, but they tell us nothing about the cost and time required to accomplish the goal. In cost-benefit

terms, we have not been able to relate benefit to cost. But of course, as the discussion in the last chapter indicated, this is true throughout the basic sciences, including social science. Basic science has concentrated its conscious attention on the goal of attaining a bit of the truth, "no matter what the cost." This lack of a systems approach in basic science is one reason why so much trivia gets published; it is "true" trivia, and "it never hurts to publish the truth." Of course it really does hurt, in terms of distraction of readers, printing costs, and the rest.

One great advantage of a computer program for this corner of the scientific realm is that eventually we may arrive at some guide to the systematic design of research. For example, one very puzzling matter kept intruding into our deliberations, namely, how much effort the program should put into generating "filter rules" which exclude classes of molecules (as in component 3), compared to the effort it puts into predicting some properties of a graph of a candidate and comparing the prediction with the original graph (components 6 and 7). Indeed, from a purely logical point of view, these efforts appear to be alike, but of course from the point of view of the writer of the program or a practicing researcher, they are not.

3. Who is the client? In the last chapter we described basic research as a highly introverted system in which client, decision maker, and designer are approximately the same person. In our example, however, this is not the case, because the present endeavor sets the designer apart; he is, in fact, trying to do something that one client, the community of analytic organic chemists, has not asked to have done, and, as we shall see, probably does not want to have done. As just mentioned, from a systems point of view it is not clear what this client does want.

However, in the actual circumstance there was another client, namely the National Aeronautics and Space Administration (NASA) and the public it serves. The presumption was that NASA and its public wished to know more about the surface of Mars and other planets. If an unmanned laboratory could be landed on Mars, it would be capable of scooping samples of Mars "soil and rock," and hopefully could fragment the samples in a mass spectrometer. But if the graph had to be transmitted to earth for analysis, the entire process might become quite laborious. On the other hand, if the laboratory were endowed with a modicum of chemical intelligence, it might be able to decide what it was looking at, and hence decide what its next steps should be. The measure of performance of such a laboratory therefore includes the reduction of "needless" time in the feedback loop.

One senses that the extra-chemistry clientele may have some con flicts with the clientele within chemistry.

4 and 5. The difficulties of trying to relate component performance to system performance have already been described. One must conclude that in this area of science—as in many others—there is very little that is methodical about scientific method. In the last chapter, the usual "textbook" breakdown of research was discussed; in this chapter we have seen how confusing is the interplay between the so-called steps of induction. Only in very restricted areas do logic and statistical inference, i.e., methodical reasoning, play a role.

6. These remarks bring us to the decision maker. There seems to be a remarkable similarity between the decision maker in science and the decision maker in industrial and government organizations with respect to the methodical aspect of their actions. Both are pluralistic. Both prefer, in a very strong sense of preference, that the important aspects of their decision making be nonmethodological, in the sense that no observer—including themselves—can explain to another mind (e.g., a computer programmer or model builder) how their decisions are made.

In the example of this chapter, most chemists would probably assert that the attempt to build a complete analyst into a computer program was futile. On the other hand, NASA and its public would probably regard the project as eminently sensible.

The experience of trying to implement the systems approach in various types of organization leads one to some generalizations about human behavior in systems. As I said, there is a general resistance to the methodical; but this resistance breaks down from time to time in certain areas. For example, a few years back the management of large-scale contracts was beset by the critical problem of trying to get all phases of the project into a smoother relationship. No method existed for doing this, but evidently some managers were better than others. Then some systems analysts invented two techniques, CP ("Critical Path") and PERT ("Program Evaluation and Review Technique"), which provided a methodical way of solving the managers' problem. For reasons that are certainly unclear, the methods caught on, and soon were required by many government agencies and industrial firms. What seems to happen is a spread of agreement among the decision makers that is probably akin to the spread of a rumor in a community. And just as rumor, when widely spread, brings its own conviction with it because so many agree, so does the spread of a methodical technique. I'd guess, for example, that some version of DENDRAL which generates plausible isomers of a chemical formula may also "catch on."

But the spread of agreement goes far beyond the rational. The methodical, indeed, tends to become stupid, as many misapplications of PERT, CP, and PPB ("Program Planning and Budgeting") testify. In other words, the manager's disinclination to justify explicitly why he does certain things comes back in the context of agreement: it's enough to say he does these things because everyone else agrees they ought to be done. The spread of statistical techniques in chemistry, biology, and the social sciences is another example. Much to the horror of the rational statistician, significance tests and correlations are run automatically, with no end in view except to run them.

We see emerging a tentative theory of implementation, to which we shall have to return eventually, since if the pluralist's resistance to rationality is so strong, then it is an aspect of the system that the designer needs to understand.

7 and 8. No more needs to be said about the designer and his intentions, which in this case were closely related to those of NASA and its public.

9. Needless to say, the designers cannot prove the stability of the system they attempted to design, either in the narrow sense that it attains a modicum of chemical intelligence by identifying molecular structures in a non-trivial fashion, or in the broader sense of the end of the last chapter—that basic science itself will survive. But the narrower sense is a very important one, i.e., to show how the battle between competing fact nets will eventually be won by one and only one net.

In the last chapter, we discussed Leibniz's approach to item 9, and his theory that all systems are alike. Thus he would view the organic chemist as being a "mind" (system) especially astute in understanding certain aspects of matter, but also as a mind that manages, does physics and social sciences, and even art now and then.

Thus when we ask, "What does the analytic organic chemist really do?" the obvious reply is that he uses chemical theory and mass spectrograph data to identify a molecule. But, letting speculation have its way, suppose one were to say that the mass spectrometer itself is teleological, that it is in fact trying very hard to do something, e.g., to minimize or maximize some function. This supposition is not altogether absurd, because in systems analysis we often look on a machine, like a weapon system, as a teleological entity, e.g., as a "kill maximizer." Now the great advantage of looking at the mass spectrometer as an "optimizer" is that we could tap the enormous and growing literature of mathematical optimization. A great many of the loosely integrated bits of chemical knowledge would then fit into place quite tightly. If we could perform such a task (and we have no idea that we can) then we would have made a first step toward an apperceptive organic chemistry. Indeed, as noted earlier in this chapter, the problem studied here is the manner in which natural forces and events can be interpreted as transformation rules which map one set of graphs, the mass spectrographs, onto acyclic topological structures representing the molecule. The speculation is that the transformation rules are based on the minimization of some characteristics of the two sets. So much for speculation.

Finally, since the illustration of this chapter concerns itself with induction, we must ask where the induction actually occurred. Induction can be defined as the inverse of deduction. Deduction is the process of using a set of assumptions to prove a theorem by some standard set of rules of inference. In other words, "Given A, infer T." Induction is then the process of finding a set of assumptions from which T follows: "Given T, find an A such that given A, one can infer T." Generally, induction does not lead to a unique solution.

Both induction and deduction in this sense occur in the design of Leibnizian fact nets and in the illustration of this chapter. Using certain theories of organic processes, we infer that certain isomers are unstable, i.e., we "deduce" that they are. Given certain sentences describing the bar graph, we seek to find other sentences in organic chemistry which imply this description, i.e., we "induce" some chemical theory.

But there is another widely held definition of induction which defines it as the process of starting with highly warranted observational statements about specific events and inferring a generalization. To understand this meaning, we must look at inquiring systems containing such highly warranted sentences. Since "highly warranted" means "well agreed upon," in this exploration we may be able to gain some insight on the elusive word "agreement" which cropped up in the tentative theory of implementation.

: 5 :

LOCKEAN INQUIRING SYSTEMS: CONSENSUS

The Search for a Warranted Beginning

There is another way to look at the discussion of the last chapters. Modern science is being forced to consider how it should plan its future because society in the shape of legislative bodies has decided to limit its resources. The natural inclination of the scientist is to plan "from within," i.e., to say what basic knowledge is and what is the best way to attain it.

The Leibnizian inquiring system provides a framework for an elaboration of the vague purpose of science to create a "storehouse of knowledge." The storehouse is a set of fact nets, gradually expanding sets of contingent truths interlinked by appropriate relationships. But the real meaning of this storehouse is the underlying principle of convergence to the absolutely true and unique fact net. Man's limited intellect can gradually raise itself to the sublime contents of the mind of God.

Today's science does not seek to modify this grand image of the scientific enterprise: most scientists believe that they are seeking to improve the accuracy of man's image of nature, and that there is an accurate image which is being approximated by fact and theory. But much needs to be done to specify more carefully how we are to plot the course of approximation, lest we waste eons in exploring fact nets which lead nowhere. The next five chapters are devoted to this task. After that, we need to ask ourselves the question whether the limit of all scientific endeavor is complete or partial knowledge, and once we have raised this question we will have implicitly raised another, namely, whether complete knowledge is a sufficient value, or whether it needs to compete with other values. At this stage we will have begun the task of planning science's future "from without," i.e., of planning from the point of view of the whole system.

In this chapter we consider an attempt to regulate the input of contingent truths into the inquiring system. It is clear that something of the sort needs to be done because the human mind is capable of concocting all sorts of contingent truths, and with great imagination linking them together into fantastic fact nets (witness, for example, the great myths). Or, in the political arena, there is the endless stream of dogmatic assertions about the social world spewed out by politicians and their aides. In order to avoid unnecessary excursions into fantasy or dogma, the inquiring system needs a filter, which will accept only those contingent truths that have some face validity.

It is only natural to think of observation as the prime source of these high-quality "facts." But observation alone is not enough, because our senses are so often confused, either because of the external environment or our own imagination. Hence we must seek those observations which are the purest, i.e., the "simplest."

We have already seen how rationalism tried to use the idea of simplicity in design to establish once and for all the existence of a guarantor. Once one is assured on the basis of a simple, rational proof that the guarantor exists, one can permit the inquiring system to become as complicated as need be, knowing full well that "eventually" there will be a convergence on the true fact net. But we have seen that simplicity of the rational proof of the guarantor does not exist, and the designer cannot rely on the strategy of seeking such simple proof. This conclusion obviously depends on what we mean by simplicity; within the rationalist framework the property of being simple is to be defined in terms of some formal structure of the process of proof. There was no need for a precise definition of simplicity to establish our conclusion, since the enormity of the task of proving the existence of the guarantor is apparent once we see that it depends on knowing the most comprehensive theory of the universe.

We now wish to consider whether the idea of simplicity can be applied to those contingent truths which are produced by observation. We shall argue that simplicity in this domain is also a delusion: it is not possible to design observational simplicity into an inquiring system. The general conclusion is that simplicity by itself can never be a basis for system design.

Now the simplicity of observation is different from the simplicity of formal proof, at least on the surface. For we shall argue that observation is simple to the extent that a community of observers strongly agree about what they observe. Our thesis takes the form that there is no simple way to design such agreement, or to confirm the existence of such agreement when it is thought to occur. Again, as we shall see, no precise definition of simplicity seems required to establish this thesis.

The conclusions of the chapter, however, are far from being negative. Indeed, in his attempt to define observational simplicity, the designer of inquiring systems will be taking his first steps toward becoming social rather than "merely" logical, by asking himself how he could design a "community of minds" (the so-called Lockean community) which agree about their sensory responses to stimuli ("inputs"). From there, we shall hope to enrich the concept of agreement in order to explore its relevance in the design of inquiry.

From this point on, we shall seek to enrich the design of inquiring systems by introducing more complicated aspects of social organizations, the Lockean community being a more or less elementary example. Although we shall concern ourselves with the meaning of observation, and hence touch on the vast literature of psychology dealing with sensation, the interest here is much more on the process of establishing agreement than in the structural details of the individual observational process.

We can see that the great contributions of rationalism and empiricism were not their theories of the origin of all knowledge, but rather their way of organizing knowledge, i.e., the theories of organization of scientific endeavor.

Simple Inputs

One way to state the problem of a Leibnizian inquiring system is to ask for a system capable of distinguishing between reality and non-reality. The Leibnizian inquirer is based on the idea that no system can be successfully designed to accomplish this purpose unless it can relate all its information to the concept of the whole of reality. In order to contrast this idea of system design with the one to be discussed in this chapter, suppose we consider the problem of designing an intelligence system for a military command. A Leibnizian intelligence system would permit the input of any kind of information whatsoever, even from unreliable witnesses or from the enemy. It would build its alternative fact nets, which in effect are alternative descriptions of the world.

The basic assumption is that the nets containing false reports will eventually "shrink" in comparison with the realistic fact net. Thus no matter how the enemy sought to load the system with unreliable information, the intelligence data base would eventually sort out all the unreal scenarios and converge on the correct one. We note that in this scheme no item of information by itself carries conviction; a piece of information becomes realistic only by virtue of being tied into the largest and most reliable fact net. For example, an item of information about the status of the enemy forces might be tied to the item saying that this information was supplied by agent X or occurred in document Y. By themselves, these two linked pieces of information carry no weight, but they gain reliability to the extent that they are linked to other reports of agent X or document Y.

Of course, there is no essential reason why an enemy could not overload the intelligence system with falsifications, especially if the enemy is very powerful. But if God is on our side, then all the enemy's false clues must eventually be found out, and the true facts will emerge. In practical cases, where a country is very powerful, some kind of system guarantor might be built into the system by the designers, who would try to incorporate in the intelligence system a knowledge of the enemy and other unreliable sources, and thus detect faulty fact nets. But clearly this technique would leave the question of reliability open-ended, since the "knowledge" of the enemy is itself a fact net that may be faulty. Even very powerful nations may become dreadfully deceived, as some of the recent history of the United States in Cuba shows.

Indeed, one can readily detect the flaw in trying to solve the guarantor problem by affluence alone. One of the real weaknesses of the Leibnizian inquirer described in Chapter 2 is that it shows no discrimination, no filtering of obviously irrelevant or false data. The illustration of the last chapter showed the need to design an inquiring system which can cut down on the number of possibilities that a "blind" DENDRAL produces.

Thus the clue seems to be summed up in the concept of attention: a filtering of masses of irrelevance and falsity by selecting only those items which are warranted. The inquiring system needs to be designed to pay attention to the relevant. But the warranty problem remains, nonetheless: How can the designer assure himself that the items to which the inquiring system pays attention are the correct ones?

In the last chapter we implicitly assumed a principle that guided us in this connection—we identified experts who then told the designer the important things that he and his system should know. However, there is a fundamental weakness in this relationship, because the designer does not really understand why the experts' judgments are relevant and valid.

To begin to understand the problem of "attention," then, we should concentrate on that aspect of "learning by paying attention" where there are no experts. The suggestion is that a certain class of information has a quality that by itself makes each item relevant and reliable. The most obvious example of this type of information is direct sensory data: the military commander will have greatest confidence in what he himself sees. In such a design, the inquiring system "traces back" a piece of information to its ultimate source in experience; the final authority of any information is a direct sensation.

Although this idea of designing inquiry seems to be based on the most forceful and simple principle of common sense, it is very difficult to make it precise, especially if one has carefully studied the reflections of the rationalists. The central difficulty, of course, is the problem of guaranteeing that a simple sensation constitutes reliable information. There are also the difficulties of determining the origin of the simple inputs, as well as recognizing simplicity itself.

Design of Simple Inputs

In the exposition of an inquiring system designed with simple inputs I shall use Locke's *Essays Concerning Human Understanding* because, like Leibniz's *Monadology,* it can be interpreted as a design document. Locke describes his inquirer as a system capable of receiving inputs; "in the beginning" it has no items in memory. In effect, the Lockean inquirer is a system with various built-in processing devices of a fairly simple sort. It has no built-in preconception of the world, no a priori information about nature. It can "receive information," combine pieces of information by logical operators, and do various other things to be described. We begin by studying the process of "receiving information."

This process starts with an entity of some kind, call it X. The process then adds to X some basic properties from a list of such properties. The basic list is called by Locke the "simple sensations." In other words, the inquirer has the direct ability of asking whether any offered X has or does not have a property in the basic list. Locke's list was constructed out of the supposed "five modes of sensation," seeing, hearing, touching, smelling, and tasting. An example of a basic property derived from each mode is yellow, loud, rough, acrid, and sweet.

The manner in which the inquirer attaches labels could be designed in a number of ways. Perhaps the easiest to understand is a "pattern-recognition" program which attempts to match some image of X with a stored set of images; if the match is sufficiently close, then the label appropriate to one of the stored images is assigned to X. Of course, this design seems to forsake the "blank tablet" theory of learning, but it is possible to conceive that the inquirer also builds up its store of images from inputs. We shall want to examine this process in more detail later on.

Thus the "reception of information" in Locke's inquirer is simply the "attaching" of one or more properties from a basic list to an "entity." Note also that, as in Chapter 2, the origin of the "entity" need not be external to the inquirer.

The rest of the elementary processing of the inquirer consists of labeling compound properties. A compound is made out of the Boolean operators "and," "or," and "not." The entire elementary "experience" of the Lockean inquirer consists of a kind of tree structure, where the trunks of the trees are the elementary observations, i.e., the elementary labeling of the received entities.

Locke's inquirer is capable of observing its own processing by means of "reflection." Reflection also consists of labeling entities, but now the entity is a process of the inquirer. The simplest properties of these processes are the labeling and compounding processes of the inquirer just described. Thus the inquirer can label its own labeling process.

The Lockean inquirer can always trace backward from any label to the more elementary labels of which it is composed, and thus back to the simplest labels. It can do this unambiguously. It can therefore ask itself how a label is constructed, or whether it is elementary.

Significant Lockean Inquirers

There is no great difficulty in representing a Lockean inquirer of the sort just described, and actually machines like EPAM (Feigenbaum and Feldman, 1963) are excellent examples. EPAM, for example, can assign structural and color properties to inputs and builds tree structures out of compounds of elementary properties.

The question of design is whether a Lockean inquirer as so far described can legitimately be considered as doing anything significant. The answer might be that the Lockean system is at least a filing system that can grow its own categories. Thus every item is given an elementary label (code) which indicates the elementary properties of the item; furthermore, the use of a given label will evoke an item with a certain Boolean compounding of properties. If we now assume that Lockean inquirers have a memory, then a label will evoke the response that the associated item is stored in memory, e.g., has been observed.

Also, we can assume that if exactly the same item is received on two different occasions, the second item will arrive at the same "station" as the first, and that the inquiring system thus has the capability of recognizing that the two are identical and simply making a note of two instances of the same item having been received. Even at this stage it can be seen that the Lockean inquiring system is far more than a "blank tablet." It needs considerable processing power to enable it to store different items in different places and to recognize that two items of like kind are to be regarded as two instances of the same input.

Nevertheless, Lockean inquirers do not gain their significance by merely being filing systems of the type just described. A filing system above all must be usable, i.e., must respond properly when someone asks it for information. But nothing has been said as yet about the nature of the elementary labeling process; how do Lockean inquirers come to have common labels? For example, if one asks a child, "What color is this?" how does the child learn to respond correctly? Evidently, the correctness of the response is judged by the adults and in general by a group of "normal" observers. Hence we must somehow design what we can call a "community" of Lockean systems having the same basic set of property labels, as well as the same labels for compounds. The community becomes the basis for judging whether a specific inquiring system is responding correctly.

Thus one teaches a child to join the community by showing the child a yellow object and repeating the label "yellow." If later when the child is shown a yellow object, he repeats the label, then we feel some assurance that he has "joined" our Lockean community of inquirers.

How Can Lockean Inquirers Inquire about the Meaning of Labels?

It will be seen, however, that even this childish example demands a great deal more of an inquiring system than we have so far put into it. In the example given, when the yellow item is received, it is taken to the appropriate storage point and recognized to be a replication of a previous item. But something fantastically more complicated also takes place in that an associated entity, namely the heard word "yellow" also occurs, and the inquiring system can see that the query, "What color is this?" is in fact asking the question about the label that should be attached to the yellow item.

Now one might design a machine capable of performing the simple childish task of giving the correct labels by guaranteeing that any item of a finite set will unambiguously evoke one label from a finite set of labels. But such a machine would be only partially Lockean because it would never inquire into the semantic meaning of terms. The problem is to design a system capable of learning how other members of the Lockean community label their "received entities." The problem would not be so difficult if there were but one stimulus and all inquiring systems had an ability to recognize "oneness," but given the welter of stimuli that occur in the environment of most inquiring systems, it is very difficult to see how we can design into an inquiring system an ability to pick out exactly that stimulus to which another inquiring system is paying attention.

Another way to state the problem is to point out that Lockean inquirers have an ability to generate sentences in the indicative mood as a result of a sensory response. An inquirer senses something, and in a moment or two it outputs a sentence relevant to what it has sensed. How does this happen? That it happens in human beings has seemed so obvious to many empiricists that they have spent little philosophical effort in analyzing the process, i.e., in discovering what the inquiring system needs in order to accomplish this feat. As a result it has become an unexamined assumption of empiricism that the discourse of inquirers must be in the indicative mood, so that many puzzling problems of the language of science arise when the situation demands, say, counterfactual conditionals, or imperatives, or interrogatives. Specifically, empiricists wonder how we can ever gain cognitive insight into value judgments because they see no way to go from the "is" to the "ought."

But their wonder is itself wondrous, for apparently they have never reflected how they got into the "is" in the first place.

In any event, as we shall see, from the design point of view there is a serious question whether the discourse about sensory experience should be couched in the indicative mood rather than, say, the imperative mood. In Lockean designs of this chapter, the discourse is in the indicative mood, and we shall proceed to suggest how such discourse could be designed.

Simplicity in the Community of Lockean Inquirers

The following account of the design of "simplicity" could become quite elaborate, and in such an elaboration would make use of the history of the concepts of perception and sensation, but such an elaboration would not serve any useful purpose here. I am essentially interested in defending the thesis that the validation of simple sensations can only be designed within the context of a community of inquiring systems; in other words, "X is yellow" can be validated by an inquiring system only if there are other inquiring systems. Hence the account will be brief, but a sketch of the elaboration is to be found in the Appendix of this chapter.

It has already been suggested how a member of the community of Lockean inquiring systems receives inputs, attaches labels, and stores the results in memory. Now we could say that whenever a label is attached to an input, the inquirer simultaneously generates a sentence. A sentence in the indicative mood is created by "input X," plus a matching against a set of images, plus an instruction "label yellow," when a successful match occurs, and another instruction "store (or output) 'X is yellow.' " We may note in passing that the sentence in the indicative mood thus depends for its existence on an instruction, i.e., a sentence in the imperative mood. The "is" is a consequence of a command; the indicative mood is servant to the imperative mood.

But what is the epistemological status of the sentence in the indicative mood? Is it "true" in any reasonable sense? Or is there any reasonable sense in which such a sentence could be false? At this stage "X is yellow" is just an item stored in memory.

To see how this item can be transferred into an indicative sentence that is either true or false, suppose we explore the suggestion made above, namely, that the "truth" of the item "X is yellow" depends on agreement with the other members of the Lockean community. The process of establishing agreement is a very subtle one. The members of the community must be able to understand each other when they describe their sensory experience. How can they do this? Suppose we say that there is designed into each Lockean inquirer a finite set of "elementary labels," and that the labels of one inquirer match the labels of another inquirer in the following manner: there exists in the total set of inputs at least one item that will be labeled by exactly one label from the elementary set of labels of each inquirer. This one-to-one pairing of labels establishes the first level of agreement. But the design problem of creating sentences that are true or false is far more complicated because the inquirers have to agree on a name for each elementary label, and they must be able to use the same name when they describe more complex experiences. For further details, see the Appendix of this chapter, where an abortive attempt is made to design a Lockean community of computers; it should be emphasized that the attempt is abortive because of the kinds of restrictions this chapter imposes, and especially the terminating quality of agreement.

In effect, if the design process is successful, when one member of the community of Lockean inquiring systems queries whether a certain label is simple, it is asking its colleagues where the label belongs in the tree of terms. To the question "Why is X labeled P?" the inquirer has two possible responses. One is to say "X is P because X is P_1 and X is P_2, and anything that is both P_1 and P_2 is P" (or some similar defense based on the compound nature of P). The other choice is to say "X is P because P is simple and X is directly observed to be P." In the latter case, the label belongs at the main trunk of the tree. If the design works, all members of the community will agree as to when a label belongs at the main trunk, and they therefore "prove" that the label is simple. Hence, whenever a sensation takes place that all the community recognize as simple, then the inquirer can generate a "truthful" sentence of the form "X is P." If, by error in the design, one inquirer makes a "wrong" statement, the error can be corrected by the other members of the community of inquiring systems. Hence the inquiry into simplicity, which is fundamental in Lockean inquirers, is a community inquiry. One must conclude that Lockean inquirers gain their significance through the existence of other similarly designed inquirers. (The problem of defining "otherness" will be discussed later on.) Simplicity of input and the ability to generate truthful sentences in the indicative mood are meaningful only if there is more than one inquirer; if there is only one, simplicity and sentential truth are arbitrary, and the Lockean inquirer would seem to become a special case of the unconstrained Leibnizian. Even so, it is not clear yet why a plurality of inquirers really negates the first criterion of the Leibnizian inquirers, the need for "innate ideas," nor is it clear how the plurality avoids the need for a guarantor. At this stage in our discussion, we can conclude that a plurality of inquirers is a necessary condition for the verification of empirical truth, but not a sufficient condition. It is this conclusion about Lockean inquirers that makes later empiricism's interest in the problem of solipsism so amazing; if a Lockean inquirer entertained the notion that it was the only existing mind, it would have to give up the idea of simple sensations, and hence its whole empirical base.

We can see now what kind of constraint is imposed on the designer of Lockean inquirers once we insist that simple sensations can be transformed—via agreement—into sentences in the indicative mood. What if two members of the community cannot agree about the simplicity of a label? Presumably they cannot use the Leibnizian trick of redefining terms because simplicity is a primitive labeling. The only answer must be that all disagreements are to be resolved by the overwhelming agreement of the rest of the community. Some relaxation on so severe a requirement might be accomplished by permitting the inquirers to introduce a language of uncertainty. Even at the level of simple sensations this design step introduces new complexities.

Thus the design of a Lockean inquirer requires the design of a community of inquiring systems in which virtually all "agree" that an input is simple or not simple, and if any disagreement ever occurs, the disagreement can be removed by re-presenting the stimulus to each of the members of the community of inquiring systems, and eventually so large a majority will agree that the voice of dissent is lost. Evidence that such a design can be accomplished seems to follow from the common experience of sensory agreement among human observers with a common language and psychological attitude.

It should be noted that the concept of agreement in the design of inquiring systems is a very subtle one, and will play a central role in the subsequent chapters. In this chapter we are primarily interested in agreement as a terminus of inquiry about some aspect of nature. As we shall see, in Hegelian and Singerian inquiring systems, agreement often becomes a signal for the inquirer to probe more deeply, and not to terminate. The point is that terminating decisions based on agreements may be the wrong mode of design; the implications for forecasting, e.g., by the Delphi technique (Helmer, 1966), or political decision making, or courtroom procedures, are obvious enough.

The "Received Entity"

The description we have just given of the formation of simple sentences from simple inputs leaves a number of problems unanswered, especially the nature of the subject of the sentence *"X is P."* This subject is, presumably, the "received entity" or "input" to which the labels of the Lockean inquirer are attached. The inquiring system, for example, says that "it" is yellow, hard, and sweet. The problem is to determine what the word "it" stands for in such a sentence.

We can begin by noting that the word "it" cannot appropriately stand for a Boolean function of a set of properties. In other words, the "received entity" cannot itself be a property or a list of properties. Otherwise the sentence "It is yellow" means merely: "Yellow-and-hard-and-round is included in yellow." Such was not the intent of Locke's design because all questions could be answered purely within a Boolean algebra by an appropriate algorithm, and the inquiring system would turn out to be nothing more than a kind of Leibnizian theorem prover with an ability to apply an algorithm within a closed formal system. Hence the word "it" must stand for something outside of the list of simple or compound properties.

What the subject of a sentence in the indicative mood can mean was the problem that Locke faced in his famous discussion of the word "substance." Lockean inquirers have the ability of assigning properties to "substances," though Locke himself was completely at a loss to explain how this ability is built into them. A substance is something in which qualities "inhere," but what does this mean? It was Immanuel Kant who gave a partial answer to Locke's design problem. According to Kant, an inquiring system capable of observing the world must have a built-in space-time framework, i.e., a coordinate system and a clock. The mysterious "it" of the sentence, "It is yellow," is an individuated point or area or volume in the coordinate system at a certain time as determined by the clock. The space-time framework is not enough, however, because this framework by itself is devoid of "content." The content is supplied, according to Kant, by a "pure sensuous intuition." In other words, the observing inquiring system also has a device that tells it whether a specific space-time volume is "received" or not "received."

Kant is frustratingly vague on the design of such a device, but the problem could be solved by means of the community of Lockean inquirers already introduced to solve the design problem of simplicity. An entity is "received" during a segment of time if all members of the community recognize its reception during that segment. Presumably, a language is required to enable the members of the community to speak unambiguously about the process of "receiving inputs," and this requirement obviously complicates the design process described in the Appendix. Henceforth, we shall use the term "input" for the validated "received entity," recognizing that this term connotes a very complicated process having nothing to do with "putting something into something else." That the meaning of "input" is only vaguely understood in many systems designs is evidenced by the strong predilection to draw arrows and boxes.

It is clear that the most significant aspect of the design of an empirical inquirer depends on the design of the community of inquirers and their communication system. We therefore must discuss what Lockean inquirers talk about, and the reliability of their talk.

Reflection

As we have already said, each Lockean inquiring system has a capability of "reflection," that is, it not only receives an input but it can recognize that it has received an input. In this sense it can act upon its own activities and label them in very much the same manner as it labels received inputs. It labels processes like "sensation," "comparison," etc., and can communicate with other members of the community of inquiring systems regarding these inner reflections; there is a common agreement, for example, about whether an inner reflection is simple or compound, and all members of the inquiring systems must agree upon a code name for a given type of reflection.

The "it" in the case of an inner process, as Kant points out, is individuated by time alone (instead of space-and-time), and, as before, has an origin in a "reflective intuition," i.e., a commonly shared experience. Hence Lockean inquirers can discuss the inner processes of their own structures, and presumably can say some very reliable things about their own processes. We note, as we have before, that this so-called introspective behavior can only be transformed into reliable sentences in the indicative mood if the whole community is in firm agreement about them.

Generalization in Lockean Inquirers

Next, the community of Lockean inquirers is to be designed so as to develop a learning process, in which they attempt to generalize their experience. Supposedly, they do this by means of induction from their agreements about specific observations.

How should we design the criteria by means of which the community of Lockean inquirers can agree on the validity of general sentences? At the outset, we might feel inclined to use the same agreement rule that was used to form the sentences about simple sensations, namely that a generalization can be assumed true if the community overwhelmingly agrees. If we did this with human inquirers in western culture, we might find that the community of human Lockean inquirers is not inclined to agree about any assertions other than tautologies and simple "facts." Hence, for these inquirers our design principle would be one of a minimum of presuppositions and only very modest generalizations. Would this strategy be justified?

Suppose we consider one very simple design of induction, in order to clarify the question. As in the case of Leibnizian inquirers, Lockean inquirers have a basic logic built into them. This logic is capable of recognizing contradictions, the validity of assertions such as "every sentence is either true or false," and so on. It also includes assertions about specific, individuated objects, as we have seen. But also the logic must include what is now called the "predicate calculus" and what Aristotle called the categorical forms: "All a is b," "Some a is b," and the negations of these two assertions, "Some a is not b," and "No a is b."

Apparently these logical forms demand some extension of the grammar of the Lockean inquirer. So far, the symbol X has stood for a specific input that is individuated by a space-time reference system. X could also represent the same entity observed on two or more occasions, provided the Lockean community could agree on rules of identification. But when we say "Every X is P" we evidently imply far more than a specific entity existing over time. The sentence "Every X is P" means "received entities having such-and-such properties also have property P." In other words, the grammar of the Lockean inquirer includes classes as well as individuals and their properties. Every member of the class associated with property P will be a received input having this property; even this definition of a "class" presents some problems, but these can be solved in an explicit way by showing how the inquirer can generate sentences in one of the four forms described above.

In the simple design we are considering, the sentence "Some *a* is *b*" can be formed by scanning the class associated with property *a*, and the class associated with property *b*, and determining whether there exist two well-established sentences of the form "*X* is *a*" and "*X* is *b*." If so, then "Some *a* is *b*" can be formed. Similarly, "Some *a* is not *b*" can be formed by the inquirer if there are two well-established sentences of the form "*X* is *a*" and "*X* is not *b*."

The next problem is to provide the Lockean system with an ability to generate assertions in the other two forms, "All *a* is *b*" and "No *a* is *b*." An obvious design would be one that permits a scanning of class *a* and class *b*, and if every *a*-input is also a *b*-input, then the inquiring system forms the assertion, "All *a* is *b*." It is incredible how often this design is assumed to be the basic design of empirical inquirers, even though no human would think of using it in its pure form. The fact that a pair of dice has come up seven five times in a row usually generates an expectation of something else on the next throw; the fact that a man has been alive today is usually not taken as evidence that he will survive tomorrow; and so on. At best, the "simple inducer" works only in a certain context; induction, in fact, is always contextual, and the design of a suitable context for generating sentences of the form "All *a* is *b*" is a very subtle one, as the last chapter indicates. For the purposes of the present discussion, we shall assume that the Lockean inquirers can be designed somehow to reach an agreement on some generalized sentences.

We should note in this connection that again there is a question whether these general sentences should be in the indicative mood. Evidently the inquirer regards "This leaf is green" to be different in its reliability from "All leaves are green." The latter seems to need some qualifier, such as "but this may not be so." What language should be designed to express these doubts?

A Priori Generalizations

In addition to the appropriate language for generalizations, there is also another very subtle design problem, as Kant pointed out. The subtlety arises because of the need to design the inquirer with a clock and a spatial framework. If the inquirer is capable of fixing each input at an "address" and a time, then the inquirer contains certain generalizations "a priori," specifically the kinematical principles that govern the clock and the geometrical principles of the space. Hence Kant's famous statement that we would not observe regularity in nature had we not first put it there, and hence, too, the apparent irrelevance of Hume's doubts about the epistemological basis of cause-effect relationships in the natural world. We might note that the status of a simple property also depends on a strong generalization about the reactions of the members of the Lockean community, namely, a universal agreement. This very critical aspect of the strategy of Lockean inquirers seems largely to have been ignored in the literature of the "logic of induction."

Nonetheless, it is not at all clear how the a priori framework of the inquirer influences the kinds of generalizations that can be made. Certainly the existence of a clock seems to order the states of nature the inquirer observes, and presupposes enough a priori knowledge to predict the state of the clock itself. The design problem is to determine the extent to which these a priori commitments influence the inductions that the Lockean inquirer makes. For example, will the future be like the past? Presumably, the future states of the clock will follow "like" patterns, or else the inquirer's entire reference system might collapse. If the inquirer is designed to conduct empirical investigations with reliability in a wide range, it probably needs a system of clocks, so that defects in the timing of one can be checked by the others. If so, the "a priori" system of causal chains may become quite rich, e.g., every class of data might require a special "standard" clock. The inquirer would have an astronomical clock, a botanical clock, a psychological clock, and so on. In such a design, the future will have to imitate the past in a very strong isomorphic sense, or else the Lockean inquirer will lose its meaningfulness.

The very critical problem of the relationship between a priori generalizations and so-called empirical evaluations has been largely ignored in descriptions of scientific method because there is a strong common-sense urge to "decouple" the a priori framework from the a posteriori. Clearly all interpretations of data require strong generalizations about space, time, and agreement, but these commitments are erroneously taken to be systematically independent of the process of observing. In the language of Chapter 3, the "data-collecting" part of inquiry is taken to be separable from the rest. This is evidence that few have considered the *design* of an empirical inquirer in which the parts are not design-separable. The problem of designing a Lockean inquirer with a minimum of presupposition is one of designing a system which makes minimal assumptions to establish empirical samples, individuation, and identification.

The Strategy of Induction

We turn now to the design of inducers more complicated than the simple inducer described above. Actually, the problem of generalization as well as data will occupy our attention throughout the rest of the book, but each time with a different perspective. Here the emphasis is on inquiring systems based on agreement; once we have introduced Kant's inquirer, the problem of design becomes radically different.

We note that the problem of generalization is not different from the problem of creating "fact nets" in the Leibnizian inquirer, except for the goals of the networks. In the Leibnizian inquirer, the goal is to create a network of sentences (culled from any available source) that will take precedence over all competing networks, whereas in the Lockean inquirer, the goal is to create as large or as elegant a network as possible based solely on the basically acceptable empirical data of the Lockean community.

A number of quite plausible Lockean designs have been very thoroughly explored in the literature. In order to make the generalizing sector behave in more precise fashion, for example, one might design into it certain "measures of confidence" with respect to the assertions stored therein by assigning a certain number to a given generalization. The inquiring system might begin to develop its own self-knowledge so that it can state degrees of improvement in its generalizing sector, and might undertake to conduct inquiry in certain directions in order to increase its effectiveness in making generalizations.

A critical design question is whether a generalization with a very high degree of confidence might begin to influence the attitude of the Lockean community with respect to its elementary data. A generalizing sector might begin to acquire such a power over the basic data system if it had arrived at a generalization which proved to be valid in thousands of cases without any contextual basis of doubt, and then suddenly found a counter instance. In this event the generalizing sector might then have authority to take the counter instance and store it into a doubtful sector of the inquiring system subject to fuller investigation, even though the counter instance satisfies all the conditions of agreement of the Lockean community. Indeed, the reason why a Lockean community agrees on an input might depend on certain powerful generalizations, so that the basis of agreement is not invariant. In other words, "dynamic" Lockean inquirers might "learn" about their own simple sensations as well as other aspects of nature.

However, to design rules governing this kind of veto power in the Lockean community is no easy matter. Consider, for example, the early history of the modern theory of numbers. Fermat speculated on whether all numbers of the form $2^{2^n} + 1$ were prime, and found this to be true for $n = 1, 2, 3, 4$. Considering the difficulty of testing numbers higher than $n = 4$, a simple inducer might settle for a guess that the proposal is true for all n. But in the context of studies of prime numbers, such a guess would be foolish. For example, the quadratic form $n^2 - n + 41$ produces primes for all n from 1 to 40, but obviously fails for $n = 41$. Hence, there exists in number theory a "counter induction" that suggests that all conjectures about prime-generating forms based on instances are highly suspect. Furthermore, in mathematics one counter instance kills the generalization. Hence in any domain where the proposed evidence is a set of favorable instances, the inquirer must be designed to ask questions about the system that generates the instances. If it takes the generating system to be like a number generating system, then no veto power should be allowed to the generalizer.

A very practical design problem of the Lockean inducer arises when the assertions generated from a set of informational inputs take on a "quantitative" form, e.g., in measurements of length or weight or whatever. The inquiring system can then begin to apply the logic of mathematical analysis in n-dimensional space and the generalizing sector of the inquiring system may propose to interpolate between two observed points or extrapolate beyond the set of observed points. The problem of design is to set up the legitimate rules governing such interpolation or extrapolation beyond what the direct information provides. We note that in this case as well, contextual justification is required to avoid the obvious failures that blind interpolation or extrapolation permits. This is the reason one should suspect the adequacy of some recent computer designs that allow all kinds of correlations to be tried and accepted if they meet statistical criteria of significance.

Next there is the design problem of the inducer which can direct where the next observation is to be made. The generalizing sector may turn the attention of the Lockean inquiring system to various aspects of nature in order to increase its confidence in certain generalizations or to modify the generalizations already held. This, of course, was not a new idea, even with Locke, but it was probably J. S. Mill (1862) who first formulated the problem most clearly. Mill points out a number of alternative strategies available to the inquiring system with regard to testing its causal hypotheses and even suggests some specific designs for its activities that more fully verify or refute the hypotheses held in the generalizing sector.

Mill's analysis of the problem, and later developments, suggested that the concept of error can play a very important role in the strategy of the inducer. In our initial discussion of Lockean inquiring systems in this chapter, we have paid most attention to the so-called simple inputs which have the characteristic that every member of the Lockean community of inquiring systems agrees as to their characteristics. In this case the input could be described as virtually errorless, not in the sense that it necessarily represents reality, but in the sense that no disagreement occurs among similarly designed inquiring systems. But to constrain the generation of knowledge to simple inputs and their compounds would greatly restrict the usefulness of the inquiring system and indeed fail to incorporate the normal practices of many of the sciences which deal in quantities and comparisons. For example, if an individual tries to compare two inputs with respect to a certain standard, he often finds it difficult to arrive at a judgment that all members of the community of inquiring systems would agree upon. Nevertheless, there can be substantial agreement within a certain range, and declining agreement as the comparison becomes more difficult. In addition to this situation, there are obviously a number of very important hypotheses governing human life in which the full generalization of the form "All a is b" is not valid, but nevertheless one may arrive at a fairly well-substantiated assertion of the form "In p percent of the cases, a's are also b's."

Both the case of discrepancy among the judgments of the members of the community of inquiring systems and the case of the relative frequency of occurrence of events are to be subsumed within another kind of logic, generally referred to as probability theory, or, if the problem relates strictly to the data itself, statistical theory. Again the literature is enormous, and the discussion of alternative designs and strategies has filled many pages of mathematical journals in all intellectual circles of the world. One finds in these journals the same kind of fundamental discussions that have occurred in the literature in the philosophy of science about how much to permit the generalizing sector of the inquiring system to govern the kinds of conclusions that it reaches. Those who believe in parsimony, for example, argue that the generalizing sector should only concern itself with inductions that can be grounded in the basic data of the Lockean community; hence the use of opinion or other forms of subjective belief is inappropriate. The opposite "Bayesian" school holds that inquiring systems are capable of generating subjective inputs based in part upon analogous situations or simply subjective feeling, and that, although these may not meet the demands of agreement of the Lockean community, nonetheless they can be helpful in formulating hypotheses and in guiding the direction of the inquirer in testing the hypotheses, as well as in the final confirmation of the hypotheses. We might note that the argument between these two schools is often not very convincing or useful in designing inquiry, because both schools fail to consider how the Lockean community should reach its agreements on the so-called basic empirical data. In the case of human inquirers, the agreements that establish the objectivity of data are based on strong subjective opinions about the characteristics of the inputs; hence it's a little late in the design process to worry about the legitimacy of "subjective" vs. "objective" judgment in research strategy. The so-called "objectivists" implicitly assume that in the inquiring system the data-collecting part is separable from the other parts, and hence that its very strong opinions about the use of subjective belief are independent of any other generalizations the inquirer makes. The objectivists make this implicit assumption uncritically because as statisticians it is "not their job" to determine how the data were collected, but "merely" to analyze the data.

If we examine the problem of statistical inference purely from the design point of view, we see that we must somehow construct an adequate "language of doubt." There is no great difficulty in doing so if we design the community to agree on some conventions and if these conventions have no relevance with respect to other actions of the community. Thus an inquirer may simply label a generalization with a letter D for doubt, or may generate the sentence "The probability of event E is p" by using more or less standard rules to be found in statistical texts. But if the sentences are also the basis of action, then it is not clear that any of the conventional language of statistics is appropriate, for it does not enable the inquirer to designate what actions are appropriate. As I ride along in a car driven by a friend, I may generate an assertion, "There is a rock in the road"; I may even label this with a D or say "probably." My friend, however, will not readily understand how such an assertion is to be translated into an appropriate action. Do I mean, " . . . and therefore watch out!"? If so, how does the inquirer get from an assertion in the indicative mood to a command?

This question about the language of doubt brings us back to the problem we postponed in order to study generalizations in the Lockean community. Some might want to say that the problem of Lockean design is not to determine how to use information for action, but rather producing information in order to describe the world. Even so, what guarantees the validity of the description?

So far we have found no adequate basis for answering this question. The design of agreement seems to be entirely up to the designer. If the designer's function is separable, then he can create Lockean systems to agree in any manner he chooses. Suppose we say that Lockean inquirers are "conventional" (or arbitrary) if the basis of their agreements is a choice of the designer and depends solely on his personal values (e.g., a flip of a coin). In order to understand the design problem of non-conventional inquirers, we can begin by looking at some recent designs that are clearly conventional.

Automated Conventional Lockean Inquirers

These conventional Lockean inquirers are represented in several recent automated designs. In all of these designs, the a priori knowledge required to receive, label, individuate, and store inputs is decoupled from the generalizing sector that performs inductions; once this simplification of the design is introduced, in principle there seems to be no obstacle to automating conventional Lockean inquiring systems, because so many of the observational and inductive methods are precisely formulated in the literature of scientific method. Indeed, the careful attention paid by the Lockean scientific community to operational definitions and statistical techniques could be regarded as a prelude to the ultimate design of computerized Lockean systems. Thus a computer system can be designed that is capable of controlling an instrument and recording the instrumental change, of compounding the data records according to various kinds of statistical instructions, of testing hypotheses, of generating new observational plans on the basis of its statistical tests, of printing out carefully written reports and charts. Some preliminary development work along these lines has already occurred in microbiology, nuclear physics, experimental psychology, and other fields. One of the most interesting developments is the proposed design of an "automated" biology laboratory that will fly to Mars and other planets. We have already seen how a part of this inquiring system can receive data from a mass spectrometer, generate all possible molecular structures, and hopefully can learn principles for reducing the number of possibilities to a very few, and thence generate new tests that will maximize the chances of determining the correct molecule.

The proposals to automate portions of research activities have naturally aroused considerable comment and criticism. For example, we might want to say that automated researchers can never create "interesting" hypotheses to test. In some of the discussion on this issue, there has been a tendency to define "interesting" in terms of computer incapability, thus turning the assertion that computers cannot create new ideas into a sterile tautology. In another sense, however, the remark may be correct; automated conventional Lockean inquirers are often confined to one basic language system, which determines the simple and unanalyzable properties of their inputs. Such inquirers are apt to become unexciting to the intellectual mind simply because they are caught in one framework, and can only go on collecting data and making hypotheses in the same mode time after time. This dullness of intellectual output, of course, also occurs in human Lockean inquirers, as many of our scientific journals demonstrate. The design problem of new representations of information will be the subject matter of the next chapter.

Even though these automated inquirers may lack insight and imagination, nevertheless aren't they conducting inquiry? More specifically, aren't they conducting empirical inquiry? In other words, what is the difference between these automated conventional Lockean inquirers and the Leibnizian inquirer? The example of the automated organic chemist of Chapter 4 is an excellent one to discuss in this connection. Is organic chemistry an empirical science? Now, the input to the automated chemist we examined was information received from the fragmentation of a molecule by a mass spectrometer. But the automated chemist accepted this input without question. In the design, the sole problem is to proceed from accepted inputs to some inference about the molecule. Such a process clearly satisfies the list of requirements given on page 34 for Leibnizian inquiring systems. From a design point of view, it seems irrelevant whether there is an "input" or whether conventional Lockean inquirers simply have "innate" ideas.

Perhaps the issue can be intensified somewhat if we look at libraries, which at first blush seem to be Lockean inquiring systems.

Libraries: Lockean Inquirers?

The objective of any section of a library is to receive all relevant documents of a given kind, to store these documents in a given place, and to retrieve them without distortion. In the case of excellent libraries, the library can compare a document received in one sector with one received in another sector. Libraries also perform the basic Lockean functions of compounding and "abstracting." The code numbers and cross-reference systems of libraries correspond to basic functions that Locke felt were inherent in the human mind in its data collection.

Most libraries of the traditional "Alexandrian" type do not perform inductions, but the more active computerized inquiring systems we have just described could be conceived as a logical extension of the documentary or archival libraries of the Alexandrian type. Thus the "library" of the future may respond to a request for information by scanning its own memory of documents, and if no adequate answer is forthcoming, the library may automatically launch an empirical investigation and make suitable generalizations. It may also conduct a series of empirical studies on a continuing basis, so that its scientific encyclopedia is forever expanding. Hopefully, it will also be able to forget in a strategic manner, so that the proliferation of itemized memory does not become monstrous. Such a library would correspond to what many people would call the "systematized collection of knowledge."

Is it really accurate to think of a group of communicating libraries as a Lockean community? Apparently there is no reason why we shouldn't. Libraries agree on the basic labeling process, and hence together solve the problem of what is simple; they also decide together what constitutes an "acquisition," and thus decide what is "given." If they work well together, they can each undertake extensions of their stored information of the type discussed above.

But would we also want to say that libraries are *conventional* Lockean inquirers? That is, is their method of cataloguing, storing, indexing, and retrieving arbitrary in the sense that the designers alone decide the method? Presumably not, because the method of cataloguing must in some sense adjust to the way in which humans react to documents. Thus we might insist that the Lockean community of libraries must include some "human" inquirers.

Another way to say the same thing is that non-conventional libraries must be "objective" in their method of cataloguing, storing, and retrieving. Libraries of the type described above could just as easily file away false acquisitions as truthful ones, even though they all "agreed" to accept a "simple" acquisition, just as the community of automated chemistry inquirers could as easily analyze contrived mass spectrometer data as real data.

Thus the clue to the design of a nonconventional Lockean inquirer seems to be the inclusion of a "human" component of the system. But what particular property of humanity is required here? We need to know this even if we decide not to go all the way with computers, because the designer must clearly recognize the human capabilities he will use, just as he must recognize the hardware and software capabilities.

Human Agreement: What Is It?

To recall how we arrived at the present point, suppose we summarize the preliminary description of agreement among Lockean inquirers (see the Appendix of this chapter). First an inquirer A recognizes a received entity ("input") and labels it; it then transmits to inquirer B a message which adequately describes the labeling; B also recognizes a received entity and labels it; B receives A's message and compares what it says with the way in which B labeled the input. If the message matches B's labeling, B records this fact and transmits a message back to A that the two labelings match. A closes the loop by receiving B's message and recording agreement. If such "agreement loops" occur for all pairs of the community, then every member recognizes full agreement about the "proper" labeling of a received entity.

We have already realized that this preliminary design of an agreement test is no simple matter, but even if it could be designed, would it really establish the basis of empirical inquiry? Return to the illustration at the beginning of the chapter and suppose that we could design a set of sensors that could communicate to each other in the manner just described. What confidence should a commander have in the agreement that such a set might reach? For that matter, what attitude should he have with respect to disagreement when it occurs? If the inquirers have a Leibnizian capability, they may be speaking in different formal languages, but it may be possible to find a dictionary for each pair of inquirers such that the assertions of one become truths for the other (e.g., "When I say 'green' it's the same thing as 'yellow' in your language"). Any apparent disagreement might thus be removed by translation. Would any community of inquirers prove a satisfactory set of empirical inquirers?

At this point, suppose we study the suggestion that the essence of the nonconventional is the human quality of agreement. While a group of computers could be "tuned in" to agreement or disagreement about their inputs, as the designer wishes, no designer can capture that immense force of feeling that takes place when a group of people recognize their complete agreement about the properties of an event. The "objectivity" of their experience rests in its clear, inevitable, unchangeable character, without a hint of the conventional or arbitrary. Of course, the rational designer will want to know whether this feeling tone of the common experience was "built into" the community or comes to it unaltered from nature. After all, a group of computers can also smile and frown and otherwise reinforce each other. And a group of humans can be as silly in their common agreement as any contrived group.

We have therefore come to an important design decision, namely, whether nonconventional agreement is one of the aspects of inquiring systems that cannot be designed. But it is far too early in our explorations to reach such a conclusion. It is probably safe to say that nothing is to be found in Locke or later empiricism that provides a helpful clue. For example, Hume's intensity of impression which becomes the hallmark of a direct sensation could be designed in any manner the designer wishes and does not therefore avoid the conventional. But much still needs to be said about more elaborate designs of observation than Locke's. In Hegelian and Singerian inquirers (chapters 7 and 9), for example, the communication system between inquirers is different from the one described above. An inquirer A is said to have an experience only if some other inquirer B can observe A's having the experience, and hence can communicate its observation to other inquirers. In this design, as we shall see, the role of agreement and disagreement becomes much more subtle than its role in Lockean inquirers.

Thus it is obvious that the problem of objective evidence in the Lockean community is a problem of system design. The attempt to design conventional Lockean inquirers is an attempt to separate the community from any larger system, the conventions arising as the designers wish, but not from any principle of design of an embedding system. One can always suspect that the so-called conventions really are related to a larger system, but the relationship is suppressed. If so, then in Lockean inquirers the objectivity of empirical evidence gains its meaning only through the way in which the research community is conceived as a part of a larger system.

Cost-effectiveness of Lockean Inquirers

The same system design point can be made if we ask the question whether the effort of empirical research is worth the cost. Consider, again, the case of the library. How elaborate should the cataloguing, abstracting, and indexing process be? A "pure" answer might be that libraries are designed to live together in order to create the most elegant systematic collection of documents possible: their main function is to be able to trace the interconnections of these documents in all relevant ways, so that a given topic like molecular structure can be "traced" through the labyrinths of chemistry, biology, sociology, architecture, and so on. Never mind that only once in a century someone queries the system about this topic, for the beauty of the system is independent of the user. In the same vein, one might insist on the purity of any Lockean community: its members establish the interconnections of their trees of knowledge as they see fit. The purity of Lockean inquirers is analogous to their conventional quality: it has no objective status.

Opposed to this philosophy of pure research are the pragmatists, who wish to see the knowledge collected by inquirers related to its usefulness. They rightly point out that no group of inquirers can collect *all* the relevant data, any more than it can examine *all* the relevant interconnections. Such phrases as "thorough examination of the facts," "study of all aspects of the situation," are sheer nonsense on the face of it. An inquirer must select from a very large set of choices a very small amount that it will examine and analyze. It does this either consciously or unconsciously—either by using a design or by letting chance opportunity lead the way. A sociologist intent on studying an organization may lay out a design of his inquiry in which each step is justified in terms of the usefulness of the results, or he may simply go in and start listening and talking, letting the output tell him what to do next. Both investigators are selective, says the pragmatist, but the first controls what is selected and the second does not.

The pragmatists, however, have tended until recent years to be frustratingly vague about how one goes about designing, for example, the retrieval system of a library. In this case one wants to tell the library system to set up a retrieval system that will be most useful to the library's clientele. This means that the library system is to become a part of a larger system and not separable as the purists demand. But in what system is the library community embedded? To date, no very satisfactory answer is forthcoming. Thus, although some attention has been paid in recent years to the relevant costs and advantages of different kinds of retrieval systems, nevertheless one finds a number of disagreements depending on one's attitude with respect to the users of the system. While it is becoming quite clear that one cannot speak of the effectiveness of libraries without some detailed information about the type of user who should be able to benefit from the library, it is not at all clear who the users are or how they should behave with respect to various Lockean communities.

Now the pragmatists' point is that the satisfaction of the user in receiving information is an indication of the "effectiveness" of the information, and this effectiveness must be sufficient to "pay for" the retrieval, else the information should not be retrieved at all. The precise form of this stipulation requires a model of the system which embeds the Lockean community, and in which the cost of information vs. its utility is described. Such a model would then determine when information should be retrieved, or when new facts should be determined. But who shall design the model, and who shall decide how to apply it?

In order to avoid the problems of nonseparability, the designer of Lockean inquirers may fall back on the principle that information is always infinitely more reliable than the cost of its retrieval or discovery. Indeed, most descriptions of the empirical methodology of the sciences actually do leave out cost-effectiveness considerations entirely, Ackoff (1962) and Marschak and Radner (1971) being notable exceptions. Often the young Lockean investigator will state his reason for examining a certain problem to be one of "merely" finding out "what goes on," as though there were no costs to be compared with effectiveness. As we have said, the trouble is that libraries and young investigators obviously can't communicate or examine everything, and hence, just as obviously they do use an implicit cost-effectiveness model. Even within the Lockean community, there must be a bounded value of information.

We are entering an age when politicians and research administrators will take seriously the philosophical claims of the pragmatists. Not all exciting research will be sponsored, and a great deal that is less than exciting will be sponsored. No one knows the likely outcome of this development, but we already have reason to suspect that the political decisions are reached by embedding the research system into the wrong larger system. This is a point to which we shall return in subsequent discussion, when we consider the various social and political aspects of science.

In any event, given the simple communication pattern of Lockean inquirers, it is not possible to design anything more than a conventional mode of handling the cost of information, in which the rules are given to the inquirers and their origin is not part of the design. A richer communication system is needed if these rules are to become non-conventional.

Separability in Lockean Inquirers

In summary, we have raised the following critical design questions about the Lockean inquirers, all of them having to do with the separability of observational subsystems.

INNATE IDEAS

We see that the true difference between the design of Lockean and Leibnizian inquirers is not clear. The "little difference" suggested at the beginning of this chapter was the denial of item No. 1 in the list given on page 34; Lockean inquirers do not build on "innate ideas." This little difference became all the difference in the world to the "hard fact" empiricist, bent on avoiding the imaginary and the metaphysical. But it is so very difficult to hold on to the difference.

It was Berkeley who pointed out that the designer cannot define a non-innate idea as an idea arising from "outside" the inquirer; the problem of the correspondence of the inputs to an outside reality is a matter of irrelevance to the designer of a Lockean system. Indeed, for such systems nothing whatsoever is added by tacking on to a given input the assertion "and this corresponds with reality." In other words, from the point of view of the inquiring system, the assertion "X is P" and the assertion "X is P and this corresponds with reality" must have equal status. What the inquiring system would have to do would be to recognize that the second assertion breaks down by means of the inquiring system's analytic device into two assertions—namely, "X is P" which is handled in the normal way, and an assertion "This sentence corresponds with reality." The second sentence the inquiring system would have to regard as meaningless, since it would have built into it no prior information about reality and therefore no basis of comparing the sentence with other stored pieces of information.

For a Lockean inquiring system, the problem of the overall "guarantor" which was so critical in the Leibnizian system is replaced by the problem of the reality of its inputs, which is also a "guarantor" problem. The Lockean inquiring system without some reality guarantor becomes conventional, i.e., capable of receiving any "information" and handling it in exactly the same way as it does inputs from any other source.

COMMUNICATION

The very significant point of our discussion is that the design solution of the reality problem lies in the design of the community of inquirers, and specifically in the design of a communication system. But the design of the community requires strong a priori commitments. Hence Lockean inquiring systems do not avoid the critical problem of the guarantor described in Chapter 3, but they do pose the guarantor problem in a very specific way. The Lockean version of the guarantor problem is to design a community with an explicit formulation of its legal structure and the determination of the manner in which this legal structure influences the "shape" of the data and the generalizations of the inquirers.

INDUCTION

Unfortunately, very little attention has been paid to the design of empirical inquiry in the literature, simply because the basic design assumptions are unconsciously assumed by the human inquirer. An excellent case in point is the modern discussion of "inductive" logic. In this discussion, we are requested to imagine a man landing on a desert island who began to observe wildlife. Suppose he sees a few birds with large wings, all of which were white and had long legs. His generalizing sector might begin to speculate that all birds with large wings are white and long-legged. In many philosophical discussions we are supposed to assume that the success of his generalizing sector in this regard is measured *solely* in terms of what would happen if the explorer undertook the rather laborious task of exploring the entire island and identifying every living object upon it that was relevant to the issue. The extent to which the generalizing sector could simulate the laborious efforts of a complete empirical inquirer is supposed to be the measure of its performance. But no attention is paid to the assumption the explorer already made in being able to identify an "it" as "white," "bird," "long-legged," etc. Nor is attention paid to the language adequate to communicate his experiences, either to himself or to others. As a consequence, we are not told how the explorer guarantees the reality of his observations.

ECONOMICS OF INFORMATION

Finally, there is the point that a "complete" empirical inquiry is impossible and hence cannot be used as a basis of evaluating the results of a sample, even in theory. The sample estimate is not merely an estimate of what a complete count would reveal because a complete count cannot occur. What is observed on the island is a consequence of the strategy of the Lockean community of explorers.

The point can be made more clearly in the case of the intelligence system we discussed at the beginning of this chapter. If these systems are regarded as Lockean inquiring systems, then it would be quite ineffective if one built into the system a generalizing sector which demanded considerable information before it was permitted to extrapolate. If a suspicious speck is observed on the radarscope and is consequently stored as a piece of information, and the generalizing sector begins to react to it, the generalizing sector may simply say, "I need a thousand more such observations before coming to the conclusion that this suspicious speck is an enemy plane." By that time, of course, the inquiring system itself may cease to exist. From this point of view, then, it is incorrect to measure the effectiveness of a Lockean inquiring system by some sort of minimization of presuppositions in the generalizing sector of the system. This point of view would argue that such minimization is comparable to the attempt of the manager of a production plant to minimize costs. One very effective way to minimize costs is to shut down the plant and sell it; another effective way is to reduce labor force and reduce inventories. But the consequences of either of these two policies may be quite disastrous from a profit point of view. Accordingly, one must embed Lockean inquiring systems in larger systems. The "objectivity" of a Lockean inquirer is defined by the larger system of which it is a part.

The Problem of Representation

In the simple communication system discussed in this chapter, we have assumed that there is one "basic" way to describe the world, which all members of the Lockean community will recognize. In more general terms, there is an isomorphism of empirical language that makes the selection of one language (e.g., English) compared to another (German) a completely arbitrary matter, once the inherent ambiguities of any natural language have been removed. Thus it does not matter whether a computer uses letters or numbers or some other set of symbols to print out its stored information, as long as these can be translated into the language of the user.

Now, imagine a community of Lockean inquirers watching two persons playing ticktacktoe. What is the "basic" data? One inquirer might feel that the "ultimately simple" data is the existence of a white space in a square, or a space with an X or a space with an O. The X and O are compound sensations analyzable into simpler components, and the community all agree forcibly on whether a space is plain white or has an X or has an O. The Lockean community which inquires about the rules of ticktacktoe by observing the players might then begin to generalize on the behavior of the players by noting that one always places the same mark in a white square, that the game is over when most of the "board" is filled, and hence that they should examine a whole set of completed games to detect the common terminating characteristic. But now suppose a mathematician joins the community. For him the "basic" data are two players who are filling up a matrix. He immediately regards the symbols "X" and "O" and their position in the matrix as number selections. Hence each player is trying to construct a matrix with some specific property. He "directly observes" that the playing of ticktacktoe is a special case of m players attempting to fill in a square matrix of size n so that it satisfies some property, e.g., so that the matrix is singular, or has identical rows, and so on.

The other members of the Lockean community might now complain that the mathematician has assumed too much, i.e., has "introduced too much of his own background," especially if they didn't understand matrix algebra. "There is no guarantee," they might say, "that the X and

O are numbers, or that the board is a matrix in the mathematical sense." The mathematician replies by making two very significant points: (1) the original Lockean community surely "introduced their own background" in recognizing a "white space" as a part of the board, as well as insisting that *X* and *O* are *really* distinct symbols, or that the placing of a symbol was the only *really* relevant behavior to be observed; (2) the mathematician's mode of representing the information is far richer because it can be used to induce rules for many other games as well.

The debate suggests a new and quite fundamental design problem which will be the basis of our discussion of Kantian inquiring systems.

Appendix

The purpose of this exercise is to design two Lockean inquirers capable of agreeing on the properties of their "inputs." The design is to be restricted to the concepts used in computer programming.

We assume:

   i. a finite set X of inputs, $x_1, x_2, \ldots, x_n$
   ii. a finite set of labels Y stored in each inquirer, $y_1, y_2, \ldots, y_n$ (i.e., a finite number of ways of sorting the inputs)
   iii. a finite set Z of "outputs" from each inquirer, $z_1, z_2, \ldots, z_k$
   iv. a community of inquirers C with membership greater than one.

The design specifications are:

   1. All members of C can receive any input from X, sort it through Y, and can output any member of Z.
   2. Whenever any member of C receives an input from X, it labels it (sorts it) by one and only one member of Y.
   3. No two inputs receive the same label.
   4. Whenever an input is labeled by a member of Y, the inquirer outputs exactly one member of Z.
   5. No two distinct labeling processes generate the same output.
   6. An output z of any member of C can be an input for any other member of C.
   7. Each member of C can associate a z received from another member with one and only one input x and its corresponding label y; call the associated triplet (x, y, z) a juxtaposition.
   8. Now every inquirer generates its own internal juxtaposition,

given an x and its associated y, and z'; it can compare two juxta-positions (x, y, z) and (x', y', z') to determine whether the elements match.

9. If the internal juxtaposition and external juxtaposition match, the member of C can output a signal to this effect; call the signal "agree" (a); if the two juxtapositions do not match, it outputs "disagree" (b).

10. The two signals a and b can be associated with two juxtapositions, j_1 and j_2, to create a triple (j_1,j_2,a) or (j_1,j_2,b) which can be an input to any inquirer.

11. For a given x and y, if every juxtaposition of every member of z produces (j_1,j_2,a), then the community "agrees."

REMARKS

A. We note that the entire process depends on a satisfactory "matching" of inputs received at different times; evidently the ability to match depends on an "a priori" capability of recognizing and individuating in the inquirer that must be of a very general character. In other words, the inquirer must have a fairly complete way of "representing" its inputs, and two ways of representing inputs for comparison could easily produce differences in the matching.

B. That the design creates "real" agreement is doubtful; for one thing, if "a" and "d" were interchanged, the community would appear to be "agreeing" when actually they fully disagree.

C. Finally, the design specifications are far too restrictive for practical use, in that perfect isomorphism of input, label, and output never occurs in practice; indeed, humans "learn" to meet these requirements in simple situations only after a good deal of practice, which is based on a type of induction.

D. Hence, the "simple" sensations of Lockean inquirers are possible only if very powerful generalizations and modes of representation are built in; the simple facts of our world are the products of our fantastic imaginations.

: 6 :

KANTIAN INQUIRING SYSTEMS: REPRESENTATIONS

The Role of the A Priori

The "Transcendental Aesthetic" of Kant's *Critique of Pure Reason* is surprising enough for what it says, but even more surprising for what it leaves unsaid yet clearly implies. What it says is that any input component of the Lockean type discussed in the last chapter must presuppose a formal structure which can at least implicitly be expressed in terms of a formal language. For Kant, this meant that the existence of an input system capable of receiving data implied that the inquiring system had built into it certain a priori sciences. According to him, these are an a priori elementary geometry, arithmetic, and kinematics.

The illustration of the library in the last chapter may help clarify Kant's thesis. If those who operate the library are considered to be a part of the input device, then it is clear that the input device has built into it at least an elementary three-dimensional geometry which takes care of the placing of documents in the stacks, plus an elementary number theory which permits coding the documents and identifying their location as well as packaging them in various alternative ways, and finally an elementary kinematics which enables the operators of the library to determine when a document is in its place and when it is not. Of course, the analogy of the library to sensation, like all analogies, breaks down because the librarians usually know what is received in the incoming mail, whereas, in the case of sensation, the whole issue of "what is received" is a very subtle problem of design.

Now if we admit that the operation of the input part of the inquiring system must presuppose certain a priori sciences, then how does the inquiring system validate the assertions of these a priori sciences? Pre-

128

sumably the generalizing sector of the inquiring system cannot possibly prove by experience the assertions of the a priori arithmetic, geometry, and kinematics of the input system because these formal sciences have already been presupposed by the input system itself. As we have pointed out in the last chapter, it follows that since the input sector of the inquirer has built-in generalizations that enable it to receive inputs, generalization in the inquirer cannot take place solely by means of empirical induction.

Kant himself seems to have believed that the basic axioms of the a priori science of the inquirer are necessary in the sense that if any of them are denied, then the inquirer is not capable of receiving inputs. Although Kant is frustratingly vague about what is received by the inquiring system, he is clear enough in setting down what he believes to be the basic formal requirements which the inquirer must have in order adequately to receive inputs.

But once Kant decides that the axioms of the a priori science are necessary, then he has also decided something else: the inquirer is capable of discovering what is necessary in order for it to receive inputs in an intelligible way. Hence, what is not said explicitly in the "Transcendental Aesthetic," but which is implied by it, is that the inquiring system is capable of examining its own methodology of receiving inputs and of discovering the presuppositions underlying this methodology if it so chooses. This process of self-examination is essential if the inquirer is to validate the axioms of its a priori sciences.

And yet how successful is this self-examination? For example, the input structure of a Kantian inquiring system requires an a priori arithmetic, because unless it can count, it cannot receive information. In the case of computers this requirement is so obvious as to be hardly worth mentioning. But there is also the design question of the kind of an arithmetic the input structure must have in order to receive inputs. Now Kant himself does not devote any attention to alternative arithmetics because in his day it appeared as though the only adequate arithmetic for an inquiring system was that set forth by Euclid centuries before in the *Elements*. But since Kant's time alternative arithmetics and alternative geometries have been developed. How does the self-examination of the inquirer determine the correct arithmetic and geometry? The fact that one cannot find an answer to this question in Kant's writings might suggest that there is a serious gap in Kantian epistemology, a gap for which Kant himself was hardly responsible, but one that constitutes a fundamental weakness in his approach to the design of inquiring systems

Most of those who have tried to defend the basic ideas of Kantian philosophy have insisted that the inquiring mind must presuppose *some* arithmetic and *some* geometry, and that therefore the basic thesis of Kant remains. However, once one admits that alternative arithmetics are possible and yet arithmetic itself is an a priori science, then one is faced again with the task of deciding which alternative among the class of alternative arithmetics is the appropriate one for the inquiring system to use. Presumably a mere self-examination of the manner in which inputs are received will not reveal the necessity of a particular a priori, or, from a design point of view, the optimality of an a priori. But since the inquirer cannot learn from experience about a priori science because it must receive experience in the framework of an a priori, there remains the puzzle of how the inquiring system could possibly decide on the correct alternative a priori science.

Two Basic Design Questions of Kantian Inquirers

From the discussion of the a priori of the inquirer in this and the last chapter, we can develop two basic design problems of the Kantian inquiring system, both arising from Kant's theory of knowledge. The first problem is to determine how the a priori structure influences the decisions of the generalizing sector of the inquirer; the second is to determine what a priori structure is appropriate, i.e., to determine the method by which the inquirer can "validate" the axiom of its a priori science. That the two problems are not separable is obvious enough, but it will be easier to discuss them in turn, and then examine the problems in more general terms.

How Does the A Priori Influence Generalizations?

Suppose we wanted to design a computer to learn the basic rules of arithmetic. How could this be done? Proceeding in the manner suggested by Locke, we might design a fairly simple generalizing sector that would take a number of arithmetical propositions and induce some rules. Since there would be a minimum of guidelines, we might expect the simple inducers to operate in a very cumbersome manner. Even learning the rules governing zero and one might be an impossible task to accomplish with accuracy. On the other hand, we must admit that the computer itself was designed by means of an arithmetical model: the designer was thinking of the operation of on and off switches in terms of a binary number system. If he could design the computer in this manner, then why shouldn't he design the rules of the number system directly into the memory of the computer? Does "learning arithmetic" make any sense as a problem for digital computers?

The reply to this design question is certainly not clear. One might say that the designer did not "use arithmetic" in his design, but rather used the theory of electronic circuits; he "interpreted" the circuits in terms of arithmetical propositions. But this response misses the point in that it does not consider the fact that arithmetic was an essential tool of the designer since it is certainly implicit in electrical engineering. The point is that in some sector of the design process arithmetic in a certain form was known. Does it make any sense, therefore, to design another sector of the inquiring system which has to go through the process of learning arithmetic? In other words, what does it mean to "learn" what is already implicitly assumed? We might note at this point that all artificial intelligence machines face this same problem: they all implicitly assume a capability that they are supposed to learn, provided we sweep in the presuppositions their designers used to build them.

As another illustration of the same theme, suppose we examine what has come to be known as "Hume's problem." In his *Treatise,* David Hume argues that the mind cannot learn the necessary connection between events by means of experience. Apparently all that experience can tell us is that an event B is preceded by an event A; there is nothing in the sense impression that reveals the necessary connection that allows us to infer that A is the cause of B. Subsequently empiricists came to adopt a skeptical attitude with regard to all casual connections between events and, as a consequence, to be especially skeptical of our ability to predict future sequences of events.

Now whether one earns the right to be skeptical in this manner depends very much on whether one has spent sufficient time analyzing the manner in which the inquiring system comes to have a sense impression. As the last chapter indicates, Kant showed that every inquiring system must have the capability of individuating what it receives as a sensuous intuition. In Kant's case individuation meant forming the sensuous intuition into a specific space-time framework. The ability to do this meant that there must be built into the inquiring system an a priori spatial framework as well as a clock. Consequently, in one sector of the inquiring system there is a set of presuppositions governing the operation of the clock. These axioms of the clock's behavior guarantee the future of the clock as well as the causal connection between the events that describe the mechanism of the clock. *Some* necessary connection between events therefore must be assumed by the inquiring system in order for it to have a sense impression at all. In what way does this commitment on the part of the input operation of the inquiring system influence the strategy of the generalizing sector? For example, should the generalizing sector take it for granted that all events in time have a causal connection simply because in the input sector of the inquiring system there is a commitment to a certain necessary connection between events?

The point is certainly not a trivial one. Galileo is observed to be trying to determine the relationship between the distance traveled by a ball rolling down a smooth inclined plane and the time required to traverse the distance. A pure empiricist of a skeptical type might feel that the entire enterprise is on shaky grounds no matter what set of results Galileo generates because there is no necessary guarantee that another try will replicate what has occurred before, however great the success of the previous tries. And yet, in order to conduct the experiment, Galileo had to build a clock. It was a water clock in which the amount of water flowing from a vessel could be controlled; the weight of the water became an estimate of the time required to traverse a given distance. Consequently, in Kantian terms, any "datum" about the rolling ball could only occur if one assumed a certain amount of accuracy in Galileo's clock, i.e., a certain determinacy of events in the mechanism of the clock. Does this commitment on Galileo's part to the possibility of designing a clock equally imply a commitment concerning the determination of events with respect to the ball rolling down the inclined plane?

Exactly the same question occurs in all of the empirical sciences. In the language of the first chapters of this book, there is a strong inclination on the part of the disciplines to separate the methodology of gathering information from the methodology of creating theories. In a science like physics where a very high degree of calibration is required, there is a very strong commitment on the part of the empirical investigator concerning the characteristics of the natural world in which he gathers his information. Very often these commitments are not transferred to the theoretical sector of the inquiring system. Thus the data-collection sector transmits numbers to the theoretical sector, but does not transmit information about the manner in which the numbers were formed nor the presuppositions that were used to form them. For example, it is typical of statistics texts to assume that the statistician starts with a set of data that are "given" and attempts to make statistical inferences from the patterns of the data; he acts as though the data were the only relevant information available.

Another example of the same problem occurs in organizational behavior. A vice president of sales wishes to forecast the sales for the coming year. For this purpose he assigns a staff to develop forecasts based on available information on past sales. If the research team adopts a separatist philosophy, it will simply assume that the records of sales made in divisions of a company are accurate and will attempt to use one of the existing methods of statistical analysis to arrive at a forecast. It will not therefore examine the presuppositions required in order to transform one of its inputs, i.e., some scratches on a piece of paper, into an intelligible record about past sales. Indeed, it will not consider its problem to be the determination of when a transaction actually becomes a sale, but will assume that this task has been solved adequately by some other sector of the organization.

But if the forecasters could use the theory that went into the determination of the sales data, their forecast might be quite different. This conclusion would be strengthened if we recognized that a past sale per se is not what interests the manager; it is rather the sales that should have occurred if proper salesmanship had been used. Hence a "forecast" is not an extrapolation of past sales, but rather (1) a determination of what could have occurred under proper sales management, and (2) a determination of what will occur under proper sales management. So put, the theory that determines the "past" will clearly be relevant to the determination of the "future."

Thus there are two opposing design philosophies of the proper relationship between the a priori sector of the inquiring system and the generalizing sector. In one philosophy, the designer will keep the two sectors operating separately with a minimum of transfer of theory, so that he can judge the effectiveness of each on its own grounds. In the second, he will attempt to look at the entire problem as though the two sectors were essentially nonseparable and thus increase the capability of both, but at the loss of simplicity of the whole design. We examine first the separatist philosophy.

Minimal A Priori Science

The first design philosophy seems clearly to have been Kant's. He attempts to find the particular set of a priori assertions that are absolutely necessary in order that an inquiring system be capable of receiving inputs. As I mentioned above, this capability can be generated provided the inquiring system presupposes a geometry, kinematics, and mechanics as well as a logic. One might mention that in the case of the Lockean inquiring system the attempt was made to reduce this list to logic alone, so that the only a priori science for the pure empiricist is the science capable of generating analytic sentences, i.e., sentences whose truth depends solely on logic.

Actually, Kant's own theory, as outlined in the "Schematism of Pure Reason" in the first *Critique,* does not seem to go far beyond the Lockean. The data gatherer must presuppose certain properties of a clock, i.e., the clock-events must obey certain laws. Hence there is an exact prediction from a given clock-event to a future clock-event. Beyond the clock, however, is the rest of reality. How does the inquirer decide that there is an objective causal connection between events that are not themselves part of the mechanism of the clock? Also, how does the clock work in assigning time to events, and finally, can the inquirer choose its own clock?

Suppose we approach these design questions in the spirit of the parsimony of the first design philosophy. The inquirer has built into it a clock which displays a series of events (e.g., it prints out the time every minute). These clock-events are causally connected in the sense that, given an event A, the inquirer can precisely predict the next event B, and conversely, given B, it can tell what directly preceded it. The event sequence of the clock is transitive and asymmetrical with respect to the relation "precedes." The remaining events of the inquirer are empirical. The inquirer can associate any such event with one and only one event in the history of the clock (e.g., a time can be printed out for every discrete input or process of the inquirer).

In this design, one might conclude that the empirical events are generated in a manner that is virtually separate from the clock-events. The fact that a white egg comes before a black bird does not provide any evidence for a causal connection between the events, nor could one predict the black bird given the white egg. At best, one can predict that a white egg at 12:00:00 will be followed, say, by a clock-event 5:00:00. In more mundane terms, those assigned to keeping the clock running have no maintenance responsibility in the generalizing empirical sector, which may develop a healthy skepticism concerning everything but the casual nature of the clock-events.

Will this separation of functions actually work? If one believes the texts written about scientific method, a large majority of methodologists seem to believe it will because, they say, it actually does work in the operations of science itself.

Since we are concerned with design in this essay rather than description, we can ask ourselves whether such a separation is desirable from a design point of view. But of course we should also ask whether sciences do in fact operate by such a separation.

There are quite a few questions we can ask about the minimal a priori design, but perhaps the most embarrassing one is: *qui custodiet custodium*—what guarantees the behavior of the clock? If the clock is calibrated to a real clock, e.g., to an astronomical clock, then the guardians of the clock sector have stored a great deal of theory. In particular, they have assumed that a set of astronomical events are causally determined. If the clock is calibrated to other clocks of the Lockean community, then how do they communicate their clock information? For example, is information about the clock sequence "conventional" in the language of the last chapter?

Finally, do inquirers operate better with one type of clock rather than another? In more general terms, can inquirers change the axioms of the clock-events, and, if so, how do they decide on such a change? Is there such a thing as the accuracy of a clock?

These questions suggest something of a paradox. On the one hand, common sense replies that one clock is better than another and that the axioms of clock-events can be changed. But how could the inquirer prove what common sense suggests is true? Can it prove these answers by experience? The paradox seems to say that it can and it cannot. Consider the famous Michaelson-Morley experiment which raised doubts about the kinematical axiom of the simultaneity of events. This classical axiom in effect stated that if two events are simultaneous for one observer they will be simultaneous for any other observer in any other place. In the special theory of relativity this axiom is replaced by the axiom that there is a maximum (finite) velocity for the transmittal of information. Common sense says that this change of an axiom of clock-events was "proven" by the experiments of the sort that Michaelson and Morley conducted. But the Michaelson-Morley experiments could not have been conducted without a clock, i.e., a set of axioms of clock-events. How could they use a set of axioms to prove that the axioms are wrong?

But the paradox may not exist after all. There is nothing strange about mathematical (Leibnizian) inquirers assuming a proposition X to prove that X is false; the method is a respectable one called *reductio ad absurdum*. But we should note a huge difference between the refutation of an empirical a priori and a formal assumption. In the latter case, we are given a formal system and wish to prove a theorem in it; to do this by *reductio ad absurdum* we attach the negation of the proposition to be proved to the formal system and deduce a contradiction. We conclude that the negation is unacceptable, and hence infer what was to be proved. But in the case of the empirical a priori, we assume an a priori axiom set and derive some unexplainable events. Then what? We certainly cannot throw out an axiom on this ground alone. At best, all that experience has shown is that something was wrong in the whole apparatus of the inquiring system but not what is wrong specifically. Indeed, experience does not even tell us that we must abandon any of our presuppositions; it just tells us (at best) that we have a problem of explanation to solve. There is no such thing as a "crucial test" in empirical science unless the designer is willing to make very strong presuppositions.

This last point can be clarified if we recall one important aspect of Leibnizian inquirers: apparent contradictions between formal systems can often be resolved by redefinition of terms. Thus in Leibnizian formal systems there is an ability to translate the terminology of one formal system into another formal system in such a way that the axioms of the first become the theorems of the second. This may be done no matter how contradictory these formal systems appear at the outset. Non-Euclidean hyperbolic plane geometry is a special case of Euclidean solid (or plane) geometry, and, vice versa, Euclidean geometry is a special case of hyperbolic solid geometry. The translations are created by defining "straight line" in a suitable manner. Hence, if a priori science is expressed in terms of axiom sets, then the axiom sets may be translated into one another by suitable definitions of terms. Thus if the inquirer's a priori structure is inadequate to explain events, a solution may be found by redefining some basic terms.

Consequently, one answer to the question of alternative a prioris is to say that there is no essential choice to be made because one alternative can be translated into another. One can define "event" and "message" so that the simultaneity of events in classical Newtonian kinematics no longer holds as an invariant; classical Newtonian mechanics by this translation technique becomes a special case of the special theory of relativity, and vice versa. Thus Kant's thesis seems to stand up even under the scrutiny of modern mathematical theory: a geometry and kinematics need to be assumed by the inquirer a priori. Which geometry or kinematics is selected is irrelevant. Can this be so? Does the mode of receiving information make no difference? Apparently it does make a difference, if the history of science is reliable on this score. The difference, some say, amounts to the difference between a simple way of handling data and a complex way. In other words, the difficulty of expressing information in one mode, e.g., by means of Newtonian kinematics, is largely removed if the information is expressed within the special theory of relativity.

Simplicity and the A Priori

The notion that a sector of an inquiring system can be judged in terms of its performance by the criterion of simplicity is not a new one in this book. It suggests that the performance of a sector of the system be judged primarily in terms of the economy of its performance at a given level of output. Presumably if the inquiring system has the ability to choose among alternative modes of structuring its information, then the best mode will be the one which reduces the time or effort spent in collecting, transmitting, and interpreting the information. The ease of interpretation is a criterion that the generalizing (or theoretical) sector uses to judge the adequacy of the information. Hence the input sector of the inquiring system must understand the requirements of the theoretical sector and adjust its mode of representing the information in terms of these requirements.

One can see something like this happening within the physical sciences where the geometries of the physicist have become far richer in terms of the number of dimensions as well as in terms of their abstract characteristics. What to the uninitiated appear to be scratches on a photographic plate, to the well-informed physicist turn out to be representations of the "paths" of electrons and other particles. The concept of the "basic data" of physics has undergone considerable change in the last three decades as the abstract mode of representation has been introduced.

But if simplicity is the basic criterion by which the input sector decides on the mode of representation of the information, then how is the sector to determine whether it has been successful in this regard? Indeed, as we shall argue later on, to make economic values like simplicity the basis of design is comparable to falling into the fire because these values are so difficult to measure, given our present knowledge of economics and society.

The question of simplicity has taken us to the second design question mentioned earlier, namely, the whole structure of the a priori science. It also reminds us that there is an alternative answer to the a priori of the input sector, namely, that the a priori assumptions might be far richer, especially if economies can be attained thereby. Indeed, simplicity and parsimony may be partially conflicting objectives. In order to consider the problem of the size and structure of the a priori, suppose we turn to the consideration raised at the end of the last chapter.

Problem Solution and the A Priori

We are watching a Lockean inquiring system trying to determine the basic rules underlying the play of a game by observing the behavior of the players. If we assume that the Lockean community operates in the manner which Locke himself described, then it will create information in which the world has various shapes and colors because colors and shapes are "simple" sensations. If the game is checkers, the individual players would be described as moving colored objects about a flat surface separated into colored squares. A Lockean inquiring system would then begin to store elementary types of information describing the specific movements of the players. We might note at this point, as we did at the end of the last chapter, that such a Lockean system has already developed a kind of strategy with respect to what it does observe and what it does not. That it pays attention to certain movements of the players and ignores others indicates that implicit in its input sector are certain guiding rules about what should be observed. Even so, a Lockean inquiring system of this type might have to struggle long and hard to determine by means of its generalizing sector what rules govern the two players and what their objectives are.

But now imagine that the observer of these two players, although unfamiliar with checkers, is very familiar with chess. He sees "directly" what is relevant; the board has become a chessboard, and the pieces are not chess pieces but are all of the same type. Such an observer sifts out a great deal of the irrelevancy, e.g., the conversation between the players, the time between moves, the behavior of other people, and so on. The inquiring system that knows chess would receive only those inputs from the movements of the two players that were deemed relevant once the game was taken to be a kind of chess game. The very first move would be interpreted as a bishop move, and the inquirer would start by "representing" the game as a game of twelve bishops on each side. The inquirer would receive very few surprises. It would soon guess by simple induction that the bishops are constrained to forward, one-place moves; it would see that the rule of "taking" is modified, and so on.

It is interesting to reflect on the nature of the phenomenal world of the two inquiring systems. For the first, the world is made up of colors and shapes, whereas for the second, the world is made up of moves of a "chess-like" quality. Which of these worlds is more basic or realistic?

But this is not the end of the possibilities. If the mathematician of the last chapter were to look at the game, he might regard the board as a matrix; movements of the players are "directly" observed to be transformations of number pairs, and hence the rules governing the play would determine the choices that each player has at his disposal in transforming available sets of number pairs one at a time into another number pair. The phenomenal world of such a mathematician is made up of numbers and transformations.

We now note that the mode of representation of information seems strongly to influence the success or failure of the inquirer in arriving at a solution. Every puzzle maker knows this to be true. For example, there is a whole class of puzzles having to do with the placement of items on a bounded surface; the players alternately place their pieces so that they do not overlap or stick out over the edge. Which player will be able to place his piece last? For one type of inquirer the search for the right answer might entail hours of laborious trial and error of alternative strategies. For another who uses a geometrical representation, the problem might appear trivial. He would see the problem as one of symmetry and know immediately from the shape of the area what the solution must be. Thus if the area is a circle, the first player will place his piece exactly in the center, and subsequently place every piece in a symmetrically opposite position to that chosen by his opponent. In this way the first player must inevitably win, since he will surely find the space to place his piece if his opponent does. The triviality in this case, however, is the result of the mode of representation. One cannot say that the problem itself is trivial, but one can say that, once a successful mode of representation has been found, the rest of the problem becomes simple to solve. Many a pure mathematician will characterize an applied problem as "trivial," forgetting the very non-trivial process by which he was trained to represent problems in a certain mode.

We can see in these examples the direct relevance of the problem of the representation of information to the problem of the a priori science of the input sector discussed earlier and to the criterion of simplicity. The input sector is to develop a strategy of representing information which will minimize the effort of the generalizing sector to solve its problems. Consider, for example, an inquiring system designed to recognize patterns, i.e., to recognize the underlying basis for the generation of sets of symbols in sequence. If I present the sequence, "2, 3, 4, 5," most schoolboys when asked for the next number will reply by presenting the number "6," having represented the sequence as the set of positive integers. A mathematician with a slightly richer mode of representation might say that the next number is "7," having represented the sequence as the prime numbers and their powers. If now we present a sequence such as "A A B A A B B" and ask the schoolboy for the next symbol or for the generating scheme of the system, he may become quite puzzled and attempt to develop certain hypotheses in an unfamiliar setting. He would not know, for example, whether the next symbol is "A" or "B." He might represent the sequence as "double A's" followed first by one B, then two B's, three B's, etc.: " A A B A A B B A A B B B . . ." Or he might represent the sequence as "double A's" followed by an odd number of B's: "A A B A A B B B A A B B B B B . . ." He might feel that the first representation is simpler, however, and this might be the basis on which he would make his final decision. Imagine, however, an inquiring system very familiar with a vast number of different patterns. It might have the capability of sifting through alternative patterns and succeed in representing its information in such a way that it can sort out unlikely candidates.

From this discussion we can discern two important aspects of problem solving. In the first aspect the problem solver is searching for a pathway that leads from the given to the solution; any pathway he eventually finds will be considered "satisfactory." In the second aspect the problem solver seeks to formulate the "given" in such a way that the pathway from the given to the solution will be the easiest one to find. In formal science a great deal of attention has been given to the creation of algorithms and other techniques that are bound to reveal a pathway from the given to the solution if sufficient time is spent on the problem. But formal sciences have not been particularly successful to date in determining how to represent a problem so that a minimal pathway, so to speak, will be readily revealed.

One can add the comment that "intelligence," as measured, say, by intelligence tests, tends to describe an individual's richness of representations rather than any general mental powers. A bright middle-class youngster may have no trouble finding out how to get seven gallons of water using a five- and a four-gallon container, while a bright black youngster may have no trouble finding out how to disappear off a street when trouble begins.

Maximal A Priori

However the optimal representation problem is to be solved, it seems clear that a certain richness of the a priori science is called for. If "minimal pathway" to a solution means economies of problem solving, then economy and parsimony are really conflicting objectives. Indeed, the use of parsimony or simplicity in design is analogous to the use of cost reduction in the design of industrial organizations. The reduction of costs per se does not produce overall economies, because the marginal cost may be far less than the marginal gain. Just so, parsimony in the presuppositions of the inquirer may not produce overall effectiveness because the relaxation of parsimony may permit a much more adequate mode of representation. The analogy of parsimony and cost reduction is itself a mode of representing inquiry as a production system. Of course, like all analogies, this one too "breaks down" under critical scrutiny; for example, parsimony in the inquirer is supposed to minimize "bias," and the representation of bias as a production concept is not very helpful. But the concept of bias suggests, as in Lockean inquirers, some fundamental, unbiased mode of representation, relative to which everything else is an analogy. Suppose, instead, we were to say that there is *no* "basic" mode of representation in the design of the input sector of the inquirer, and that a maximum flexibility in representation is desirable. Hence, instead of attempting to minimize the influence of the a priori of the inquiring system on the information, the designer should try to maximize this influence in order to represent the information in a manner which facilitates problem solution.

To illustrate this idea, imagine an inquiring system with the elementary capability described in the last chapter, i.e., an ability to receive inputs and to structure them in an elementary way and to compound them and generalize on them in the manner of a Lockean inquirer. Imagine also that this inquiring system has a component in which is stored a set of models. Any given model can be conceived initially as having primitives, axioms, rules of derivation, and theorems in accordance with at least a partially well-formalized language. In addition, each model has the capability of taking any input and transforming it into a sentence of the model. In other words, any stream of inputs becomes in this inquiring system a sequence of sentences within one or more of the models that are stored in the a priori component. In the example of the play of a game given above, the input stream might be represented in any one of the following sequences: (1) "Player A pushes a red piece one diagonal square (model is kinematical, with shapes and colors)"; (2) "Player A moves his bishop one move (model is chess)"; (3) "Player A transforms the number pair (2, 3) to (1, 4) (model is number theory)."

Now imagine that the inquiring system has an "executive" which examines whether a given input is appropriately interpreted by a given model. At the simplest level such an executive would ask whether the input stream as represented within a given model seems to be leading rather easily to an appropriate solution. If the answer to this question is affirmative, then the inquiring system will continue to transform the input into the model, i.e., to look on the world through the "spectacles" provided by the model.

From this discussion we can discern at least five critical design problems from inquiring systems with maximal a priori models.

1. What input should the inquiring system process? That is, if we make the realistic assumption that many inputs are being received, then there must be a design economy in selecting a subset of the inputs for scrutiny. How shall the inquiring system sift out exactly the relevant inputs to process through its models?

2. How should an input be translated within the language of a model? For example, there are many ways of looking at a problem from the viewpoint of number matrices. Consider the following game. There are nine slips of paper on which are written the numbers from 1 up to 9; each player is to select one of the pieces of paper in turn; as soon as a player has three pieces totaling to 15, he wins. What is the optimal strategy? If one translates this game to the game of selecting numbers from a three-by-three "magic square" whose rows, columns, and diagonals all add up to 15, then the game is "seen" as a game of ticktacktoe, and the solution becomes trivial. How did the inquiring system decide

to translate the input into a magic-square matrix rather than some other? Indeed, many writers have suggested that finding a rich analogy is a creative process that cannot be analyzed, i.e., presumably cannot be designed.

3. How does the executive decide to judge whether the translated inputs provide a good basis for solution? In most of the examples cited so far, the solution appears either obvious or very difficult, so that the executive would presumably have an easy time of it. In most real situations, however, considerable detailed examination of the problem would be required within each model in order to determine whether the model is or is not an appropriate framework for problem solution. Perhaps the most difficult aspect of this question is the way in which the models are designed. One suggestion is as follows. The input is given to a model as some symbol sequence. The model takes this unintelligible sequence and "translates" it into a sentence of the model language. The model then determines if the sentence is true. If so, it so informs the executive and proceeds to the next input. If not, it tries another translation; if this "fails" it so informs the executive, who stores the sentence in a Leibnizian fact net, using some fairly simple ("basic") language. The fact net generates simple rules by using very elementary inductions; the fact net could then be called the "simple inducer." The rules of the simple inducer are then transmitted to the models for translation and test. If a model does poorly, e.g., fails to translate three or four sentences, it is dropped by the executive, i.e., it is considered a fruitless analogy. The design can be enriched by permitting the models to query the input sector, e.g., by asking whether instances of their axioms are true. In this case, the axiomatic form of the model becomes a critical design strategy. See the Appendix of this chapter for a more detailed outline of such a design.

4. How is the executive to judge whether a solution has occurred? In most of the examples cited so far, the problems are framed in the style of a "puzzle." Puzzles are mental exercises concocted so that one model or way of thinking, is *the* appropriate pathway to the solution. As a consequence, solutions are easy to identify when they occur, as is the appropriate model. In the richer problems of everyday life, one rarely finds that the problem can be construed as a puzzle, and one of the most difficult aspects of realistic problem solving is the determination of whether or not a solution has occurred. Indeed, our earlier references to systems and subsystems indicate that a solution in the sense of a separable optimum does not occur in real systems. Another way to say the same thing

is that in social problems like pollution and poverty, there is no authorized source for terminating the inquiry, as was supposed in the last chapter.

5. Is the whole idea of the design of a maximal a priori as given above appropriate? The design suggested above and in the Appendix is one in which the input generates some "elementary" structure, e.g., a string of symbols or shapes of some kind, which are then translated into the language of the models in the a priori sector. Here the executive has no control over the elementary input structure. In other words, "what is received" is not a decision of the inquiring system itself. Is this an appropriate way to regard the Kantian inquiring system? Certainly Kant's own discussion of "pure sensuous intuition" seems to imply that he would accept such a philosophical basis for the design. *What* experience is telling us is very difficult to ascertain, because we always have to shape the "messages" into some a priori form to make them meaningful. But, for Kant, experience does contain information that is "given" to us and which the inquirer cannot change. Otherwise the Kantian inquirer seems to become a special case of the Leibnizian: all the knowledge is implicitly within the inquirer. To escape the clutches of the Leibnizian design, most students of problem solving accept the thesis that some aspect of the input stream is not under the control of the executive. In Chapter 4, the chemistry inquirer was represented as an examination of a tree structure. Essentially the task is to determine whether at a given stage it is worth going on to examine the rest of the branches of the tree or whether one should, so to speak, give up a whole branch as being a less than likely method of solution. But the basic form of the symbol stream that is processed in this manner is "given" to the inquirer.

What Is "Given"?

We are now in the position to form what Kant called an antinomy—an apparently irreconcilable clash of theses—between the Leibnizian and the Locke-Kantian theories of design. The antinomy can best be phrased in terms of the design of the Kantian inquirer suggested above:

Thesis (Leibnizian): All aspects of the input stream are in principle under the control of the executive.

Antithesis (Locke and Kant): Some aspects of the input stream are not under the control of the executive.

In effect, the antinomy is a further elaboration of the first design principle of Leibnizian inquirers: the need for innate ideas. In defense of the *thesis* is the argument that any attempt to define a "given," i.e., an uncontrollable aspect of the inquirer, always results in an unintelligible design. Kant's own writings seem evidence of this point. He says that inquiring systems have a capability of receiving inputs, i.e., a receptivity, and that what is received is a pure sensuous intuition that is shaped by the forms of space and time and made intelligible by the categories as they are described in Kant's *Transcendental Logic*. Consequently, the inquiring system is incapable of recognizing in an intelligent way what it is that begins the process because "recognition" implies a structuring which apparently does not occur in the purely sensuous intuition. This is exactly the same problem that occurred in the last chapter. There we saw that one could, if one wished, simply regard the origin of the process to be given arbitrarily, thereby creating "conventional Lockean inquiring systems." But these conventional Lockean systems are special cases of Leibnizian inquiring systems, i.e., their conventions are controlled by an "executive." In the case of the Kantian inquiring system, the problem of the need to put all control in the hands of the executive is even more acute, because we see that it is not possible to describe the given in any linguistic form: linguistic forms already confound the pure given with the sensory and intellectual forms of the a priori.

In defense of the antithesis is the obvious need of the inquirer to "learn about the world," i.e., to become objective. The failure of Leibnizian inquirers to find an internal proof of the convergence of all fact nets to one world picture implies the need for an external source of information. If there is no "given" in experience, then there is no difference between deduction and induction. Induction merely becomes the search for symbolic strings of symbols that lead from one sector of a fact net to another according to rules; but this is exactly the manner in which deduction is described. Induction should represent the effort of the inquirer to form a comprehensive picture from what is given to it. Otherwise, the whole of inquiry is merely a logical exercise.

The Relativity of the Given as a Strategy

One clue to the solution of this antinomy is to be found in our earlier discussion of the separability or nonseparability of (1) the assumption-making of the a priori sector from (2) the generalizations of the inquiring system. In all of our discussions to date we have looked on the designer of the inquiring system as in some sense separate from the inquiring system itself. Specifically, in the case of the design of problem-solving inquirers, it is clear that the designer considers himself to be "above" the situation that the inquirer itself is trying to study. The designer "puts in" the given and examines how the inquiring system goes about trying to solve the problem. In effect, the process is much longer than a process beginning with the "givens" and ending with the inquirer's solution. The process goes back to the designer himself who has originated the givens. How did the designer decide what in fact to give the inquirer to solve? If we could understand the process by which the problem concocter makes up the problems, perhaps we could begin to see our way to solving the antinomy. We would, in fact, have a theory about the problem concocter. If we knew enough about him we would know which of his inputs we should process, the likely way in which to translate these inputs into the language of a model, and which model is likely to provide us with a solution. Also, knowing him well, we would know how to judge the quality of the solution in terms of his desires. What we would then mean by a "given" would be relative to the problem originator.

Indeed, if we sweep in the problem concocter, we may be in reasonably good shape to answer some of the questions about the a priori that are listed on pages 142–144 above. The executive will learn about the favorite models of the problem poser, and will explore these first. He will learn what satisfies the problem poser, and thus learn when to stop. And so on.

How will this suggestion work when the problem poser is reality? Can we design an executive to "learn" about the favorite models of nature? Does such an idea make any sense at all beyond the obvious requirement that the inquiring system learn the secrets of the real world? In the case of problem posing and problem solving, the inquirer may learn to sweep the problem concocter into its world image, but what does it mean to "sweep in" reality?

Problem Solvers Do Not Exist Unless They Are Observed

A richer solution of the antinomy is needed. E. A. Singer in his *Experience and Reflection* "turns the tables" on the empiricists by showing that the reality of an observing mind depends on its being observed, just as the reality of any aspect of the world depends upon observation. The suggestion is that what is "given" to an inquiring system is a problem of another inquiring system observing the first in its problem-solving activities. The "given" is a concept of the observing inquiring system. It is no wonder then that designers of inquiring systems like Locke and Kant, who separated themselves from the system that they were observing, ran into severe problems in trying to describe the nature of the given. Once we bring Locke and Kant into their own inquiring systems, then we begin to see in what sense the word "given" should be taken. But this is a story of its own that unfolded during the nineteenth century. It is a story to which we shall devote our attention in the next chapter.

Appendix

A schematic design of an inquiring system is capable of representing information in alternative ways. The design is described in terms of phases, each phase being an enrichment of its predecessor.

PHASE I

1. An input sector which produces "basic" sentences, e.g., sentences that ascribe a set of "basic" properties to individuated items.

2. A set of model sectors; in each model there is:

a. a translation subsector that can take the basic sentences of the input sector and translate them in one specific way into sentences of the model;

b. a list of assertions that are true in the model (these are not necessarily the axioms of the model, but rather display the most pertinent aspects of the model);

c. an ability to determine whether a translated sentence from the input sector implies, or is implied, by a sentence in the model list. More generally, the model sector will build a Leibnizian fact net, using the input sentences and the model sentences, and will determine the extent to which this fact net is "satisfactory" according to some criterion.

3. An executive who can:

a. accept problems to be solved;

b. initiate or turn off the exploration in any model sector, according to general criteria of satisfactory exploration;

c. take certain sentences of the satisfactory fact nets of the model sectors and store them in a common language as tentative partial solutions of a problem;

d. decide when the problem is solved;

e. in the case of failure to find any satisfactory fact net for a given input sentence, can perform simple inductions in the basic language of the input sector and store these as tentative partial solutions.

PHASE II

Consists in increasing the power of the executive to modify the "basic" language of the input sector as well as the translating subsections of the models and the sequence of critical assertions of each model. In other words, this design permits the executive to use past experience to modify the linguistic structure as well as the search procedures.

PHASE III

Consists in adding information about the source of the inputs and the problem origin, so that the executive can relate his strategy to a specific input source or specific problem poser.

We note that phases II and III involve the self-reflective paradox: if the executive can "use experience" to modify the language or search, or learn about the input source or problem source, how does he represent this experience? Does he process the experience through a "super" set of models? This problem is a part of the problem of realism of the inquirer: How does the executive come to realize that his problem solutions are realistic as compared to imaginary? How does the ability to recognize the problem or input source in terms of models ever get the inquirer in touch with reality?

: 7 :

HEGELIAN INQUIRING SYSTEMS: DIALECTIC

Objectivity

Objectivity is the hallmark of all excellent inquiry, and yet its meaning remains elusive. The objectivity of a result seems to imply that no one is obliged for "external" reasons to accept the findings, that each inquirer may learn how any other inquirer conducted its objective inquiry, and each inquirer is free to test the methods used when they are objective and thus confirm or refute the results. Thus a necessary condition for objectivity is that the behavior of an inquirer be capable of being observed. But what does this requirement that one inquiring system be observable by another really mean, and how does the inquirer acquire objectivity from this process?

At the outset we should note that "objectivity" is closely related to "object" in meaning as well as in sound. Kant seemed to have thought that objectivity occurs when experience is shaped into a "general object," i.e., gains its form and intelligibility from space, time, and the categories. But even this shaping of experience is not enough, as the discussion of the last chapter shows. We also need to design into the inquirer an ability to see the "same" object from different points of view. In a sense, we made a beginning of satisfying this design requirement in the last chapter as we explored different modes of representation. But now we need to develop the additional idea of an "object" as a collection of interconnected observations in which each observer can examine how another observer views the world. The "objectivity" of experience is to be based on some kind of interconnection of observers.

Now it is almost obvious that many "points of view" are required

149

to create an "object" like an elephant or a university. What is not so obvious is that many "points of view" are also required to produce the objectivity of a property like "green" or "straight." For Lockean inquirers, all that is needed to attain the objectivity of a simple property is a strong agreement in the Lockean community. And yet, as we have seen, if there is no control on the agreement, the so-called objectivity of the property becomes no more than a convention that—in the case of computers—can be changed at will by changing the program. We seem to need a watchdog who can monitor this sort of thing and decide whether the community is conventional or not. A "watchdog" watches; i.e., the observers need to be observed in order to gain objectivity even for simple properties. In other words, no observation can become objective unless the observer is also observed objectively.

To Be a Mind Is To Be Observed

If we apply Berkeley's dictum—that an object gains its objectivity only by virtue of its being observed—to the property of being an observer, it must mean that something can only be said to observe by virtue of its being "observed to observe." This rather obvious point about inquiring systems is often neglected. Designers wish to create computers to "solve problems," "observe patterns," and so on. But whatever ability the computer attains in any of these directions becomes objectively valid only because the designers observe that the computers are functioning in a certain manner. To understand what the computer is doing objectively, it is essential to know what the designer is doing. The "fact" that computers "solve problems" is as much a description of the behavior of the designer as it is a description of the output of the computer program.

What would a philosophically astute empiricist have to say about this need for an observer-of-the-observer? One answer seems almost obvious and in effect became the cornerstone of a particular type of philosophical system developed in England. The answer says that there are actually two radically different ways in which the observer may be observed: (1) he may observe himself directly, or (2) he may be observed "inferentially" by another observer. These two ways of observing are taken to be radically different because in the first case the self-observer is assumed to be almost completely accurate about what he observes, while in the second case the "other" observer can only infer what is actually being observed "inside" the mind. "Another" observer can only observe what occurs "on the surface" or at the interface of the two inquirers, whereas the same inquirer can observe its own inner states directly.

I said that this answer to the meaning of the observer-of-the-observer is obvious, but it remains to be seen whether it is really a satisfactory design principle. Indeed, from the design point of view it already seems to involve some weaknesses, e.g., the awkward distinction between "inside" and "outside" or between "same" and "another" which is surely very difficult to define and apparently serves no very good purpose. Yet there is something very compelling about the thesis that each of us has his private thoughts, sensations, and feelings.

The Self-knowing Self: The Subjectivity Syndrome

British empiricism certainly seems to have retained throughout its long history the notion that immediate sense data and the inner pictures and images of the mind are the special property of the self, and knowable only by the self. This has given rise to what might be called the "subjectivity syndrome" of a certain popular type of philosophy both in the United States and in Great Britain. It takes the form of the assertion, "I and I alone can know the inner states of my own mind and can only *infer* the states of other minds." Accompanying this assertion are a number of corollaries, e.g., "I can never be aware of someone else's toothache" or "I can never be sure that someone sees the color green as I do." There are also some fairly serious philosophical doctrines connected with this philosophy, such as solipsism and the inability to compare utilities, which have flavored the intellectual life of a number of social science disciplines. All of these doctrines imply that because I and I alone know the inner states of my own mind, no one else can possibly supply any better evidence about my own inner states than I can; at best other people can only infer the properties of my inner states by observing my outward behavior. Also, since I have no way of developing an inquiring system that reaches beyond my own observations, it follows that I have no evidence of the independent reality of other minds.

According to the subjectivists, because I and I alone know exactly what a toothache feels like to me and because I have no way of comparing the sensations that other people have except in terms of the grimaces and other signs of distress which they show and which are alike to mine, to infer from these observations of outward behavior the existence of a pain similar to mine would be to go far beyond what the evidence itself supplies. It would permit the generalizing component of the inquiring system to make a leap in the dark on the basis of one instance alone, namely, my own sensations.

Subjectivism is a very weak philosophy with very strong implications. Toothaches may be matters of minor concern in the total history of humanity, but one important implication of the subjectivist doctrine is that it is impossible to compare the values of two or more members of a society other than in terms of their simple preference ordering. So convinced were economists that the intercomparison of preferences must be excluded that there occurred in economic literature a number of serious attempts to provide a basis of social choice which is free of the need to compare the preferences of two separate individuals, except in an ordinal sense. Thus an observer can look at a person and see that he chooses X rather than Y. Such behavior occurs at the interface of the observer and the observed person. But, according to subjectivism, he cannot observe how the person felt "inside" when he made the choice; specifically, he cannot observe the intensity of the person's preference, although the person clearly does have an inner feeling and is aware of it himself. The intensity of preference is taken to be a subjective evaluation which cannot be communicated to another.

The economist's reluctance to assume cardinal utilities and richer forms of measuring human values has had its influence on the entire theory of optimization in system design, especially when the "system" involves human beings. For some, the only legitimate "optimal" is a "Pareto optimal," which is often a very unsatisfactory criterion in the design process.

Now all sciences recognize difficulties in the design of calibration in measurement, calibration being basically the method of communicating a method of measuring. Calibration is never perfect, so that errors inevitably occur. In length measures, for example, one must try to relate a method that uses a yardstick to a method that uses a micrometer; nevertheless, the comparisons of units are made with "reasonable success." What is it, therefore, that prevents a like comparison of units in human value measurements? The answer is that the "real" unit is encased in the individual person, and there is no way of laying one person's unit "alongside" another's; in other words, the "fundamental" mode of "direct" comparison of units is supposed to be ruled out in value measurements.

"Direct" vs. "Inferential" Observation

All these versions of solipsism and subjectivity arise from the assumption that the observer must play a peculiar and separate role in inquiry. In order to be sure that the observer is safeguarded, so to speak, one designs all inquiry as an emanation from a central and effectively unanalyzable set of "direct," "fundamental" operations of each inquirer. The fundamental operations cannot be compared or observed because once one permits such comparisons, the authority of the central observer disappears and one is apparently left with no basis for objectivity.

The design principle of subjectivism has many important consequences, all based on "levels" or "degrees" of knowledge. The central core attains "direct" knowledge, and the further one travels from this core, the more doubtful are the assertions the inquiring system makes. For example, there is no way in which the pure empirical inquiring system can seriously consider the task of predicting future events. It must simply regard its own data as its sole type of reality, and when it speaks about the future, it can only do so in a kind of poetic fashion. The modern subjectivist shies away from concepts of forecasting and in general from the whole notion of the redesign of systems in terms of their improvement, since the term "improvement" itself implies an ability to forecast the future.

Again, there is the recent distinction which occurs in game theory between the concepts of "uncertainty" and "risk." The risky situation is taken to be one about which the inquiring system can make probability statements which are based on directly observable events. An uncertain situation, on the other hand, arises, say, in a two-person game where the actions of the one player depend on his own developed strategies and these strategies cannot be predicted from the relative frequency of past plays, since they are based in part on the one player's concept of how the other player will conduct the game. It is said to be impossible for the one player to obtain evidence about the "inner states," i.e., the strategies, of the other player.

Subjective empiricism in philosophy, psychology, and economics has never undertaken to defend its fundamental doctrine or even to state it clearly, perhaps because the doctrine seems so obvious and reasonable. The phrase that frequently recurs in Locke's *Essays*, "If one will but look into his own mind," is simply reiterated down through the decades as a perfectly satisfactory approach to philosophical reflection and a knowledge of one's own mind.

Personal Knowledge and Community Knowledge

The doctrine no doubt seems obvious because it is difficult to understand how one inquiring system can be a direct observer of another's internal states and processes. For the Lockean community, it is enough to establish an isomorphic agreement, so that the same inputs are followed by the same verbal expressions and by expressions signifying a common agreement; the exact matching of internal processes is not feasible or even essential.

But the subjectivist not only excludes direct observation by another of a subject's state of mind, but postulates a maximum accuracy on the part of the subject with respect to its own direct observations. When one inquirer observes what impression it had as a result of a simple input, the answer it gives must be essentially correct. In other words, subjectivism introduces the distinction between personal knowledge and community knowledge into the design of inquirers.

Community knowledge implies careful control and scrutiny on the part of other inquirers; personal knowledge does not. Thus if a scientist undertakes to create knowledge for the community, he must write down what he intends to do, and then, if he is a careful scientist, he must keep a log of what he has done, and finally he is obliged in his reports to present his findings in such a way that any colleague can, if he wishes, observe exactly what the scientist has been doing. Failure to comply with any of these conditions is a defect in the procedures of the Lockean inquiring system. While it must be admitted that science often tends to be rather careless about applying these three conditions, nevertheless in all cases of dispute the Lockean critic of a scientific endeavor has a right to call for further and deeper explanation of what has been done so that he can replicate the work of another scientist. Thus the overt behavior of a scientist must itself be subject to observation by other scientists whenever the purpose is to create community (common) knowledge.

The basic trouble with subjectivism's distinction between personal and community knowledge from the designer's point of view is simply that it leaves no room for an explicit design. Furthermore, the designer will argue that it is not at all clear why we should ever try to design such a distinction into the inquirer. Of course, the nondesigner will reply that each individual wishes to preserve his own subjective feelings as a part of his world. Thus while the designer is not apt to remove belief in personal knowledge by any kind of logical argument, his point is that the manner in which the subjectivist introduces personal knowledge into the design of inquiry seems altogether wrong. He wants to give personal knowledge of sensation the status of highest accuracy, and he goes on to link this personal knowledge in a very tenuous fashion with community knowledge. He thinks that personal knowledge of one's own sensations creates types of behavior that other inquirers can observe and transform into their own personal knowledge. This seems to require a very awkward type of design, and all the awkwardness can be removed simply by removing the need for an "emanation" of knowledge from a central core. If "personal knowledge" of one's own sensations simply means that inquirer A observes itself as an object in exactly the same way that another inquirer B observes A, then all the "mystery" of internal states disappears. In such a design the designer would not lose control at what is one of the most critical points in the whole activity of inquiring systems, namely, at the point where information is received and interpreted. If the observer can be observed, then he who observes someone "feeling a toothache" can learn what is being sensed at least as well as the one who "has the pain." In human inquirers, in fact, there seems to be ample justification for asserting that one person may be far more sensitive to another's reactions than the other is to himself.

What would result from giving up the supremacy of privacy is a complete revision of the restrictions on empirical inquiry. The so-called "basic" or "fundamental" units and comparisons of the inquirer are no longer basic or fundamental from the point of view of another inquirer; preferences, direct sensations, and the like are the output of all inquirers and have no special epistemological status. In Hegel's terms, the "immediacy" of sense data becomes a mediating concept of the reflective mind.

Consider, for example, the comparison of personal preferences or utilities. Instead of merely asking each person in a community to state his preferences as he observes them, one would also determine how each person observes the preferences of others. If there is a fight over the allocation of resources, not only do A and B state their requirements, but B states what A wants and A states what B wants. There is no a priori weighting of these observations of individual needs.

Representations of Observational Behavior

The design of an inquiring system we shall consider is based on the principle that inquirer A's information about inquirer B's internal states may be as reliable as B's own reflections about his internal states. But what does it mean to say that one inquirer can observe another's states of mind? One answer is to be found in the discussions of the last chapter: we can extend the Kantian theme of representation to the observations of the states of mind of an inquirer.

One mode of representation has already been described: one can observe a chemist as he examines various items under the microscope and see that he writes down his results in his log and carries on various other kinds of activities. This mode of representing the chemist can be enriched by comparing the activities of two chemists engaged in essentially the same type of work. One can make statistical analyses of the results and represent differences in their behavior in terms of their personality types, the laboratory social environment, and whatnot. This way of representing observation takes the observers to be psychological individuals, with describable psychological properties: motivation, sensory response, oral behavior, etc.

Another way of representing the observer is to consider him as a physical entity. The observer is now regarded as a physical input-output device that receives impressions at its surface and transmits these via its neural system to some central core, where the "message units" are stored and retrieved. By this method of describing the sensory organs and neural structure of the human observer one can check whether or not certain distortions are introduced either by the external instruments or by the particular physiological structure of the observer himself. Hence we can say that some properties of the stored inputs are products of the instruments that the scientist uses, or that some properties are the product of a particular neurological structure of the observer. We can develop some ideas about the relation of inputs to the real objects of the world outside; the "objective" inputs are those that are not distorted by the instruments or internal transmitting processes.

Objectivity in a Physical Description of Mind

Now when the observer represents the inquiring system as a physical object responding in a physical manner to physical stimuli, the subjective empiricist may feel that the life of the inquirer has been taken away, since there is no representation of the "inner states," as he feels them to exist. Nevertheless, if we accept this mode of representation for the moment, we can see that the design problem of objectivity can be described in a precise manner. The problem is to determine whether an inquirer's account of a situation is "objective." The observer of the inquirer can see both the object (stimulus) and the "inner state" of the inquirer as it is represented by a physical description of the stored inputs of the inquirer. Suppose the observer is able to classify the stimuli into identifiable elements, as well as classify the "inner states" of the inquirer. If he then observes that for each stimulus property there corresponds one and only one inner state, he could say that the inquirer is responding "objectively." If, furthermore, the inquirer is also observed to output a set of symbols that are in one-to-one correspondence with the inner states, then the inquirer is reporting its experiences objectively.

This description of objectivity should be compared with the design of agreement in the Lockean community; in the design of agreement (see Appendix of Chapter 5), the community of inquirers could not observe the inner states of the other members, although in order to describe their behavior we did construct their inner worlds. As a consequence, the community cannot tell whether the set of stimuli is mapped onto the same set of internal states in each inquirer, even though we had to pull ourselves aside and do precisely this. Also the community members cannot really tell what the word "agree" means, except that it occurs for each inquirer when associated in a certain manner with a stimulus. Finally, each member of the community can only know how he reacts, and hence has no objective knowledge of a stimulus "outside" his own internal states of mind.

But if the adjective "objective" refers to a certain type of experience of an observer of the inquirer, then all the mysteries of the subjective empiricist vanish. An object is "there" because it is a part of the experience of the observer-of-the-inquiring-system and is observed by him to have a certain relationship to the internal states of the inquirer. To the observer, the object is "outside" the inquirer, and the observer can precisely determine whether two inquirers agree. He can even determine whether they see the color green in the same manner. Thus the mystery of Kant's sensuous intuition vanishes: it was after all nothing other than a construct in Kant's mind as he observed the human inquirer.

Suppose we say, as Hegel did, that the process by which one mind observes another is self-reflection (or self-consciousness), recognizing that this old-fashioned term is both practical and common in its meaning here. Managerial control in a firm is a self-conscious process, as are the controls of scientific, traffic, and educational systems.

In order to keep the characters clear, while not intending to impute any specific meaning to their roles, suppose we call the mind that is being observed the "subject" and the observing mind the "observer." The inquiring system we shall examine is the "observer-of-the-subject."

The Problem of Objectivity in General

A number of questions come to mind as soon as we gain the insight that objectivity is a property of an observer of a subject, i.e., a property of self-reflection. First of all, we see that the "subjective" has not been eliminated at all, because in the act of self-reflection the objectivity ascribed to the subject becomes itself a subjective state of mind of the observer. *That* a subject is having an objective experience is a subjective experience of the observing mind. All we have said so far is that for a total observer-of-a-subject to be objective, it must have a component capable of receiving inputs about the manner in which another component receives and processes inputs. Self-reflection is a necessary condition for objectivity, but not a sufficient one. The observer of the subject may be totally wrong about what it observes or reports. As we shall see, we have still to understand how objectivity can be established as a fact about the inquirer.

Next, the nature of the "internal state" of the subject is not clear. So far we have represented this internal state in physical terms, as though the observer, for example, were describing a human subject in much the same manner as he observes the innards of a computer. But the whole point of subjective philosophy is that any such representation of a person's sensations and feelings is largely irrelevant as far as the "actual" internal state is concerned. Can we represent how a person "actually" feels or what he sees when he observes the deep green of pine against the intense blue of sky? As we have said, this may be the Kantian question applied to the understanding of a subjective state of mind: What are all the representations of the subject that make up its reality? One recalls the story of the blind men trying to describe an elephant while each touched some part of the body. But in the story there is a wiser observer-of-the-subject who "sees" the elephant for what it "really" is and can therefore laugh at the antics of the blind. But who sees for the community of inquirers? Who can tell us when our modes of representation leave out crucial features?

Our design question therefore is this: under what circumstances does a set of representations of an object capture the essence of the object, i.e., become objective in the most general sense? Perhaps subjective philosophy is universally correct: no *thing* can ever be adequately represented by an image of its nature, in which case subjective internal states are not different from any other states of nature as far as comprehension of their essential nature is concerned.

The Mechanist Theory of Objectivity: "Information"

Whenever philosophers feel called upon to describe a whole class or universe, their tendency is to search for a dichotomy that will serve as a beginning for a more elaborate classification. This dichotomy should produce new ways of looking at old problems. The dichotomy that seems to be called for here is the dichotomy between the mechanical and the teleological—between observations that are taken to be reactions to a stimulus or message, and observations that are taken to serve some purpose. The essence of "mechanical" observation is alienation: the observed subject is opposed to the observer. Either the subject is passive and the observer active, or else the observer receives "inputs" (and hence is passive) while the subject creates "outputs" (and hence is active). The observed and the observer cannot be the same mind, and must be two opposing aspects of a process. The alienation is well known in experiments in which humans (or other living beings) are subjects, or in interviews in which the behavior and attitudes of people are being studied. The experimenter or interviewer is the observer and is a different kind of person from the subjects. He is supposed to have no prejudices, to be rational, to be completely honest in his reporting, not to care who is right, and so on, while the subjects are interesting only because they have prejudices, are irrational, dishonest, self-seeking, etc.

The flavor of the opposition between the observer and the subject seems to be well captured by the term "information." The inquirer is "formed" by a certain type of input, much as a computer is formed by a program. Hence the "information" that is stored in an inquirer is taken to be the set of all reactions of the inquirer to inputs of a certain type. Specifically, we imagine an observer-of-the-subject who can identify an input as an accurate sentence that describes some aspect of the natural world. If this sentence is received and stored by the subject, then the subject has reliable information. The mechanist theory of information goes on to say that a "state of the world" is simply a conjunction of sentences about the properties of objects in the world. The mechanist has an answer to the question: What set of representations capture the essence of an object? The set is comprised of all sentences that accurately describe the object, i.e., all sentences that ascribe all the correct properties to the object.

Information: Master and Slave

The judge of the accuracy of information is the observer-of-the-subject, who in some way holds a dominating role, because the accuracy of information for the observer is taken to be independent of the wishes or purposes of the subject. Information for the mechanist is there, and what it says cannot be changed by the subject. We determine whether a subject is objective by determining whether its stored information corresponds with reality; if it does, the information is factual ("objective"). If it is factual, it cannot change, no matter how the inquiring mind may change: "a fact is a fact" is not a tautology, but rather a statement of an hypothesis about the relationship between information and mind.

According to the mechanist hypothesis, the fact dominates the subject simply because he has no choice about the facts. He may wish ever so ardently that men love their fellow men, but when the facts reveal that men hate each other instead, then the inquirer must bow to the authority of fact. In this mechanist relationship, man becomes the slave of the master who is information, or, rather, the slave of the observer-of-the-subject, because the "world" of the inquirer is a creation of an observer-of-the-subject. It is a world that the subject cannot change once the observer has fixed it in his mind. More precisely, there is some way of observing-the-subject in which the past states of the world of the subject can be accurately ascertained by the observer, and hence are not changeable by any action on the part of the subject. The mechanist hypothesis states that the past as constructed by an "accurate" observer-of-the-subject is unalterable once the states of the world are correctly ascertained. Here is true alienation of self and fact; the self is the slave of the master fact created by "another" observer.

Once the mechanist hypothesis is stated in these terms it appears almost absurd. What master observer-of-the-subject can ever gain the authority to "fix" the facts, i.e., to legislate what is information and what is not? Such a legislating mind puts each man in bondage to a mysterious and unknown master.

Information: The Conquering Lord

And yet the mechanist hypothesis about the nature of objectivity has infiltrated practically every aspect of intellectual and social life. Government information agencies consider themselves to be recipients and storers of various "pieces of information" of interest to the citizen and to those who serve the citizen in military and non-military capacities. The information is fixed, and cannot, under penalty of the law, be altered. Our whole theory of statistics is based on the notion that pieces of information can be numbered and represented symbolically (x_1, y_1, etc.) and that the task of the statistician is one of aggregating these "given" bits of information in various ways.

In logic, too, the mechanist philosophy has had its strong influence. The logician is primarily interested in sentences, and he has come to think of sentences in terms of their semantic content. Any given sentence that is not a tautology may express a factual description of the world, and it is the task of the semanticist to put the factual aspects together in a pattern that will be useful to the reader.

Finally, in that part of mathematics called "information theory," information is reduced to common units, and specifically in the case of digital information, to "bits" of information. An important problem for many information theorists is to extend the concepts developed in information theory relating to redundancy, etc., to useful concepts of meaningfulness. So fascinating are the developments of information theory that many writers speculate on the possibility of completely describing the human brain in terms of an information-processing device. To do so, there needs to be another mind that can accurately observe what the brain is "really" like, although proponents of this physical reductionism do not tell us how this "other mind" can also be a mechanical information-processing device. How does it happen that in all these widely accepted approaches to information, there is the master observer-of-the-subject who has the authority to fix the unalterable status of information in various "data banks" of government, industry, etc.?

The Subject As the Willing Slave

Somehow it must be the subject himself who delegates this authority to the master observer-of-the-subject, because our tradition states that we have freed ourselves from dogma. But if we examine the reasons why people accept facts, it is not at all apparent what policy decisions are being followed. For example, we accept information because:

1. It is provided by "experts" (physicists, doctors, engineers, etc.)
2. It is produced by a highly bureaucratized system with "built-in" controls (accounting systems, registration systems, etc.)
3. It is such that no one feels inclined to disagree (current state of the weather, existence of a fire or war, etc.).

In each of these instances, who is the master observer-of-the-subject? Who says that the information of experts should be accepted? Who says that bureaucratic information is reliable? Who says that facts that no one disputes are accurate? The answers are readily at hand. The authority of the expert arises out of the recognition he has gained from his peers. If you want to know whether Jones is an expert, ask Smith, who is also an expert. The people, i.e., the "subjects" in our earlier terminology, decide who these guardians of expertness really are. The authority of the bureaucratized system arises out of the acceptance of the system by "auditors." If you want to know whether a company has kept its books correctly, ask an auditing firm to check on their procedures. In the last case—universal agreement—the master observer-of-the-subject seems to be a "collective mind"—a mind that is "more than" all the individual subjects and that can pass judgment on what each individual says.

The designer of inquiring systems is less interested in whether the master is the expert, the auditor, or the collective mind than he is in the basic design principle that justifies each of these choices. Why should Smith be accepted as an expert on the expert advice of Jones? Why do we accept the findings of auditors, or let the collective mind dominate our notion of what is really happening? Perhaps some hint of the answers to these questions can be found by examining a similar list where the subject does not recognize a master:

1. In moral matters there are no experts
2. In accounts of the saving of souls or the blessings conferred by philanthropy or federal aid there are no auditors
3. On the true causes of war and poverty, there is no collective agreement.

It is interesting to note that this second list seems far more important than the first. The experts can tell us "facts" but they can't tell us what our ultimate values should be. The auditors can guarantee the statement of "assets" of a company, but they can tell us nothing about the social value of these assets. The collective mind can agree that a calamity has occurred but can say nothing about why it occurred.

The Subject As a Manager

This comment on the second list suggests another basis for a policy of the subject that will govern the master observer-of-the-subject: instead of mere blind willingness, the subject should delegate authority whenever the net benefits warrant doing so. The net benefits are made up of two components, the gross benefit of the policy and the cost of carrying out the policy. Thus we would all like to check on the advice of an expert, but it would cost us years of education and training to do so, and hence the maximum net benefit accrues when we trust the experts. The experts may be wrong on occasion, and this is added into the cost of the policy. Even so, the net benefit may be maximal when we trust them. But on matters of ultimate value, the net benefits are not maximized because the experts disagree or the subject does not know how to identify an "expert," and thus the costs of trust are too high. Such a policy makes the subject a "manager," who permits the master observer to rule whenever the net benefits so decide.

The net-benefit basis of information policy is what I have previously called the "teleological" approach to information because it emphasizes purpose (means and ends). The teleological approach appears to reduce the alienation between observer and observed. If the subject is forced or blindly willing to accept the facts about the world because of the dicta of an observer-of-the-subject, the alienation is severe. But if it is the subject after all who uses teleological considerations to appoint the master observer-of-the-subject, then the alienation seems to disappear. The "facts" are, after all, the creations of the subject's own policy making. Thus when a simple sentence of the form, "This is green," is stated, one subject may respond, "Is it indeed?" and the other may respond, "So what?" The "Is it indeed?" response is the response of a blindly willing and alienated subject. Such an inquirer receives the offered piece of information and stores it as a piece of information in the mosaic of bits of information in its memory. It accepts the legislation of a master observer. On the other hand, the response, "So what?" albeit rude, seems to be a freeing response. Here the subject is in no mood to receive unless the offered piece of information can be perceived as useful in some plan of its operation.

The Paradox of Teleological Information

But it must be apparent that the teleological basis of information policy utterly fails to solve the problem of authority, nor does it really remove the alienation of subject and observer-of-the-subject. All it does is to suggest a new question: What are the costs and benefits of trusting the master? Who establishes the evidence, pro and con, for answering this question?

But how difficult a question is it? Are the facts fairly obvious, so that the collective mind of agreement could be created to solve them? Consider a rather simple item of information, e.g., information about a constraint on the behavior of the receiver. For example, the information is in the form, "There is a log across the road in front of us." If the subject is in the mood of "So what?" he may very well drive on ahead and pass over the rotten log without interruption of his normal course of action. In this case the offered information is of no value whatsoever; indeed, it does not even stand as a "fact" in the receiver's world. In other words, in the case of the "So what?" attitude, something becomes information only because it can find a justifiable place in the *total* scheme of the subject's activities in such a way that its position in the total scheme of things, as William James (1911) puts it, "makes a difference" in what the subject actually does. It begins to look as though the question of net benefit even in simple cases is not a very obvious one at all.

In order to explore the net-benefit policy of information more precisely, we can discuss in more detail the illustration of non-separable systems in Chapter 3, the control of inventories. An inventory system is simply a system which stores items that have use at various points of time. The reason for manufacturing and storing the items ahead of time lies in the extreme inconvenience that may occur in trying to create an item at exactly the point where the need arises. This is a familiar enough situation to any householder who habitually stores various sorts of canned goods, sugar, salt, and the like, thereby incurring an expenditure ahead of the actual occasion of the need simply to avoid the enormous inconvenience of obtaining the items from the store at exactly the times when they are needed.

Now what is the relevant information that anyone who tries to set up an inventory should have? Well, first of all, he should have some estimate of the inconvenience that may be incurred when items from inventory are demanded and are not available. This will guide him in the relative importance of storing various types of material in his inventory. Next, he should know how much it costs him in time and money to procure items for inventory. Third, it would certainly be worthwhile if the person controlling the inventory had some knowledge of when the needs were apt to arise. This would enable him to plan his inventory storage policies over time in accordance with these needs. Finally, the controller of inventory should have information about the costs of holding items in inventory over long periods of time. For example, he may find that certain items deteriorate, or that the purchase of an item ahead of time prevents his making use of a more up-to-date item when the need really occurs. He may also find that his capital is tied up in inventory and is not available for other opportunities when they occur. In certain countries government taxation policies will also impose penalties on him for holding items in inventory.

There may be other types of information important to the holder of inventories. He may wish to determine, for example, how long it takes to receive an item into inventory once an order has been placed, and he may want to have some general information about the quality of the items in inventory and whether they really meet the requirements for the items when the needs occur. Suppose, however, we restrict ourselves to the four elementary aspects mentioned above, namely, the cost of shortage and of placing items in inventory, the demand for items from inventory, and the cost of holding an item in inventory over a period of time. All of these are examples of teleological information.

From the point of view of an inquiring system the problem now is to determine these "basic" pieces of information. If we were to adopt a purely empiricist approach to the problem, we would ask ourselves what we must observe in order to arrive at suitable information concerning the four basic questions. What, for example, shall we observe when we are asked, "What does it cost to order and place an item in inventory?"

The natural reaction to this question would be to say that one should look at the past history of the inventory system. What, in fact, *has* it cost in the past to place an order and receive an item into inventory? We might therefore examine the activities of the people who place the order and begin to add up the cost components in terms of labor and materials required to initiate the order and to process it. We would do the best we could to develop a careful description of the exact way in which orders are placed and the kinds of controls that are imposed, and would try then to relate these to actual cost output on the part of the entire system. In this regard, the inquiring system would adopt an "Is it indeed?" or passive role in its opposition to nature. A state of nature would be a conjunction of assertions about how orders have been placed as determined by an expert master observer. Similarly, we would try to determine what losses had occurred in the past as a result of shortage.

In the same vein, if we were to tackle the problem of requirements from the demand point of view, we would try to search through past records to determine at what particular points of time requests for items had been made, in what form they had come, how large the quantities were, and so on. The "objective" demand on inventory is something that is told to the inquirer, who receives what is given, i.e., the data, in a passive mode.

Finally, turning to the question of the cost of holding items in inventory, we might conduct a careful search of past records that give evidence concerning obsolescence, deterioration, taxation, and the like. To estimate the cost of capital tied up in inventory, we would try to estimate the extent of demand the inventory system places on the available capital of the corporation or agency; we would then try to estimate the actual "lost opportunity" cost of this tie-up of capital.

This "Is it indeed?" approach to the inventory problem is the one most often followed by operations researchers and others who try to assist management in the control of inventories, but from the point of view of the designer, the whole procedure seems very weak. For example, the particular inventory system may rely on a certain resource of supply. This resource may require that a certain number of days' notice be provided and that the inventory system must pay a certain penalty each time an emergency order is placed, and so on. It would naturally occur to the designer of the inventory system to ask whether or not the "given" source for the inventory is appropriate. Perhaps if the inventory system itself could control its own source, a number of the penalties associated with replacing items in inventory would not occur, and at least the total cost of placing regular orders and emergency orders could be vastly reduced.

If this were the case, it would be simply incorrect to say that the *relevant* information about the cost of ordering and placing items in inventory is to be found in the practices of the existing resource agency. Anyone who confined his attention to this kind of information would simply fail to acquire information, whatever "data" he found.

In other words, if the assertion about the cost of placing orders is in the form, "The cost has been k dollars per order," the assertion is not yet information; its opposition to the purposive inquirer is quite different from its opposition to the mechanical inquirer. In the case of the mechanical inquirer, the information will be received if it is properly authorized; in the case of the teleological inquirer, it will be used if it fits into a total plan of action.

Or, again, in the case of the demand for items from inventory, the designer may find that the person asking for items from inventory does so according to a certain convenient pattern from his point of view, but has no real need for the item when the requests are made. The designer may in fact discover that if the persons making the requests are rewarded in certain ways, they can smooth out their requirement schedule so as to avoid almost all of the emergency situations that have occurred in the past. Thus, shortage costs and demand patterns are not passively received by the designer.

Finally, when we consider the problem of holding items in inventory, we may discover that in the past the organization has often failed to take advantage of opportunities to use capital most profitably. In this case, descriptive sentences about past opportunity policies would not constitute information about the cost of tied-up capital in view of the fact that were these policies improved, the actual costs of tying up capital in inventory might be considerably greater than one would estimate from a description of past behavior.

All this amounts to saying that the inventory system is embedded in a much larger system. The theme, of course, is merely an application of the theory of systems developed in Chapter 3. An inventory system is a non-separable part of the rest of the system, and "information" about the characteristics of the inventory system from the teleological point of view depends upon the way in which the total system is viewed. The cost of ordering and placing an item in inventory is not an isolated "piece" of information; a cost-datum contains within it a picture of what the entire system is like, just as do the requirement schedule and the costs of holding the items in inventory. The inquirer cannot passively observe what the costs and requirements of an inventory system are. He must infer what they are from a total picture of the entire system in which the inventory system is embedded. Each cost factor in effect is a mirror of the entire system: it reflects the way in which the entire system works so as to generate a certain penalty associated with a given type of action that is adopted by part of the system. Thus the sentence, "The cost of doing x is k dollars," is an abbreviation of the sentence, "The entire relevant system has such-and-such properties among which is the cost of doing x."

To recapitulate, there are two radically different ways of defining "observation." Mechanical observation is defined as a "reaction" to a stimulus: an inquiring system A "observes" an object X if another inquirer B observes that A is "reacting" in some manner to X. The reaction may be the flash of a neuron or the flick of an eye or a spoken word or a string of symbols. Once the observer-of-the-subject observes the completed process of stimulus and response, then for him the subject has "observed." To know that a subject has observed "objectively," we need the authority of a master observer-of-the-subject. Teleological observation, on the other hand, is a way of observing the world so that the resulting information is useful to a purposive being. To know that a subject has observed "objectively" we need to know the total system in which the subject acts. We can justify the appointment of the master observer-of-the-subject by means of a teleological argument, i.e., the master is the appointed servant of the teleological subject. But this justification simply complicates the relationship because the subject cannot decide without teleological information, and yet he cannot acquire *objective* teleological information without knowing the whole system.

The Search for Objectivity: Infinite Regress or Vicious Circle?

The pathway to objectivity seems to be either an infinite regress or a vicious circle. It would be an infinite regress if the designer were always to evoke a new master observer to legislate over the old master and his subject ("Jones is an expert because Smith says he is, and Smith is an expert on Jones' expertness because Brown says he is and . . ."). It would be a vicious circle if the designer were to permit the subject to appoint the master and the master to appoint the subject ("Jones is an expert because Smith says he is, and Smith is an expert because Jones vouches for him.").

It is interesting to note that the regress is merely called infinite, while the circle is called vicious, even though the circle appears to be the more innocuous of the two. From now on, these two characters will play their role in the design of inquiring systems; the problem is either to design a regress of inquirers that will somehow collectively approximate objectivity, or to create a circle that is not vicious. In effect, the Lockean community is an attempt to build a non-vicious circle, because each member's objectivity is guaranteed by the agreement of everyone else. In political designs, Locke's is a system of "checks and balances," but as in the case of the Lockean community, it is not apparent why agreements of the interested parties constitute the objectivity of their beliefs.

Information and Weltanschauungen

For the present, we turn our attention to the possibility of designing an infinite regress of inquirers that stands for more than a simple and dull, "A is right because B says so, B is right because C says so, etc." A teleological inquiring system wishes to know whether a piece of information is correct. In order to decide on this matter, it creates an image of its world—a *Weltanschauung*—that provides one picture of the inquirer's alternative actions and hence the relevance of the information and the way it should be used. For example, the inquirer wants to travel. from X to Y. One *Weltanschauung* says that there are four means of travel, and provides the times and costs of each. On this basis, the inquirer selects the conveyance requiring minimum time and cost. Convenience and safety, according to this *Weltanschauung,* are irrelevant, i.e., no matter how risky the travel, the "optimal" conveyance remains the same. Another *Weltanschauung* may say that the "real" objective of the inquirer is not to travel, but to communicate with a distant colleague. The picture of the set of alternative actions shifts, and the interpretation of the "data" of the first *Weltanschauung* becomes quite different. The "cost" of travel must now include lost opportunities of the inquirer to use the travel time for other purposes.

Hence in the teleological theory of information, the sentence, "X is a piece of information," is valid only when embedded in a certain *Weltanschauung,* i.e., way of viewing the entire system. It follows that an inquirer attains objective information only if he chooses the correct *Weltanschauung.* But this conclusion seems to leave the whole problem of design up in the air, for where is the master observer who can accurately determine the characteristics of the relevant world of the decision maker?

The Hegelian Dialectic

The historical solution to this question was first suggested by Kant and later elaborated by the post-Kantian German philosophers, and especially Hegel. In the "Transcendental Dialectic" of his first *Critique,* Kant considers some classical hypotheses about the origin of the universe, its boundaries, and the immortality of the soul and its freedom. He presents side by side two equally compelling arguments, each based on all the facts and reasons his ingenious mind could find. One argument shows convincingly that the world could have had no beginning in time, while the other shows with equal conviction that it must have had a beginning. In the same vein, one argument rationally proves that the world is bounded, and the "antithetical" argument demonstrates that it is not; the thesis proves the immortality of the soul, the antithesis proves the mortality; the thesis proves that the will is free, the antithesis that it is not. The point of all these exercises is to establish Kant's grander "synthesis"—that unconstrained reason leads to contradictory conclusions because it is permitted to go beyond its proper use as a coordinator of sensuous inputs into the inquiring system.

In Hegel the Kantian design is made more explicit. First, the inquirer must be endowed with a richness of experience. In Hegel's philosophy, this meant exposure of the mind to a vast array of psychic events, in literature, history, philosophy, and science. In the more mundane approach to the design of inquiring systems, the requirement might be interpreted as a loading of "information" in the mechanical sense discussed above, where an attempt is made to acquire as broad a sweep of the "data" as possible. The same idea is familiar to anyone who has tried to study organizations with a view to improving them; the first few months may be spent in "looking at and listening to" as many aspects as possible.

Next, the inquirer must generate a conviction about some fundamental thesis; it must have the capability of believing wholeheartedly that a certain point of view is correct. For Hegel, this conviction must be rooted in a strong feeling as well as the kind of logical demonstration Kant provides in the "Transcendental Dialectic." It is essential for Hegel that the inquirer live its conviction as well as think it because the conviction for Hegel is a stage in the psychic development of the inquiring mind.

It goes without saying that this requirement is vague from the point of view of a thinking-type designer who wants his requirements to be explicit. A formal approximation to the "living reality" that Hegel talks about might be accomplished as follows:

The designer undertakes to construct a "case" for a point of view, in effect a defense of a thesis "A." The design of the case for A constructs a *Weltanschauung* in such a way that the wealth of "information" in the inquirer's data bank is interpreted by means of the *Weltanschauung* to support thesis A over all other possibilities.

What the designer tries to do is to reverse the usual design procedure of data-to-model-to-optimal. Instead, he starts with the optimality of a policy as a "datum" and then constructs a view of the world in which certain data become relevant (information), other data become irrelevant (non-information), and the relevant data plus the plausible world view maximally support policy A. In other words, he proceeds from optimal to model to data. The formalization of this design procedure can be sketched as follows. The inquirer has a data bank of elements, d_1, d_2, . . . d_n. The elements of the data bank are symbols of various kinds. They may be numbers, graphs, mathematical equations, reports, etc. In the example of the inventory problem, they would be the collection of all the things that the operations research team has heard about costs, products, personnel, profits, etc. No "datum" by itself has any epistemological status, i.e., it does not say anything about the world.

The inquiring system also has a set of formal models which can each be interpreted as a description of the "whole system," W_1, W_2, . . . W_n, and the interpretations of which are non-identical. Any datum d_i conjoined to a *Weltanschauung* W_j implies one or more items of information. An item of information does have epistemological status, i.e., is teleologically meaningful. In other words, an item of information has the property that it can be used as evidence relative to a thesis A. The thesis A is an assertion not contained in either the information or the *Weltanschauung*. By "use as evidence" is meant that the item of information lends a certain positive, zero, or negative credence to the thesis (as opposed to an item that is couched in a symbolism that has no meaning relative to the thesis).

The formal design of Hegel's "living conviction" is to select a *Weltanschauung* that maximizes the credence in the thesis A, i.e., a W_0 which when conjoined with each of the elements of the data bank produces an information set that maximizes the evidence for A. In plainer language, the inquirer sets about showing that there is a way to look at reality so that the data can be interpreted to support the thesis.

In this account of the Hegelian design, it is not clear where the thesis comes from. How does a person acquire a conviction, or how does the inquiring system select the thesis whose credence is to be maximized?

Leaving this question unanswered for the moment, we introduce the next character in Hegel's drama. This is an observer-of-the-subject, who looks on the act of personal conviction "objectively." He says, "I see that this mind is utterly convinced that thesis A is true; I wonder why." The spirit of this observer is in opposition to the subject. Its "I wonder why" is the "So what?" mentioned earlier. The opposing mind is in the mood of "I wonder why another conviction wouldn't do just as well." In order to observe the subject, the other mind conceives another conviction and asks what it would be like to be equally convinced of this "antithesis." Such is the manner in which one Hegelian mind observes another: in the mood of opposition.

Now in Kant's "Transcendental Dialectic," the antithesis was found in a straightforward logical manner. If the thesis is, "The world had a beginning in time," the antithesis is, "The world had no beginning in time." A classical logician would want to say that either the thesis or the antithesis is true; Kant argued that both are epistemologically unprovable, and that the only "truth" to be found is that they both try to extend reason beyond its proper domain. In Hegel, on the other hand, the antithesis is not the contradictory of the thesis, but rather its deadliest enemy. It is an anti-conviction of forcefulness at least as great as the conviction. The "deadliest enemy" concept is found most clearly in politics. The deadliest enemy of democracy is not nondemocracy, but a very explicit and detailed political design called the Communist Party.

Now the very effort to maintain one's conviction in the thesis generates opportunity for the deadliest enemy. The effort to preserve democracy leads to wire-tapping and secrecy, in fact to nondemocratic policies. If a nation as a whole is convinced that dictatorship is correct, its very conviction will breed revolution. The observer-of-the-subject is not a dispassionate "other mind"; it is passionately dedicated to destruction of the subject's conviction. The very activity of observing the subject creates a mood of opposition.

If we try to design this very living drama of conflict into the inquiring system, it is not clear how we can capture its life. Following the more or less deadpan approach to the design of conviction given above, we can set down the requirement that the antithesis B be so selected that of all alternative countertheses to A, B has maximum credence. This means that there exists a *Weltanschauung* W which, when conjoined with the data, maximizes the evidence for B, and the maximum "score" for B exceeds that attained by any other counterthesis and its maximizing W.

Evidently, the meaning of this design conviction (thesis) and counter-conviction (antithesis) depends on the set of data, the set of *Weltanschauungen,* the meaning of "conjoining data to a *Weltanschauung,*" and the measure of credence, all of which are vague at this stage. But the basic design idea needs to be explored before the formal details can be made more precise.

It will be noted that in the dialectical design described above the thesis and antithesis have the same status. The antithesis is built out of the thesis by the building blocks of the data and the *Weltanschauungen*; but the thesis could just as easily have been built out of the antithesis. Hence we miss that aspect of the Hegelian picture in which the antithesis looks upon the thesis that generated it in a way that is different from the attitude of the thesis. The revolutionary looks down upon the reactionary. The reactionary in his conviction can only think that the revolutionary is crazy or criminal; he must utterly reject him as an unnatural evil or a meaningless mind. But the revolutionary understands the nature of the reactionary full well; for him the reactionary's conviction is based on a natural selfish greed and hypocrisy. As at other points in this chapter, e.g., personal vs. community knowledge, we see that the design does not seem capable of representing the living idea of the philosopher.

In the fourth act of the Hegelian drama another type of observer-of-the-subject enters, who observes not the conviction but the opposition.

He is a quite different observer-of-the-subject because he tries to see how the opposition arises out of the particular kinds of minds that clash in their convictions. But he is no compromiser of the bargaining sort that one finds in labor and international disputes. He is also in opposition, an opposition to the very nature of the conflict, and he does not seek to deal out rewards that will keep all parties reasonably content. Instead, he builds a new world view in which the nature of the conflict is understandable, but which shows that at a higher level the conflict is merely one aspect of reality and not the critical aspect. The conflict in fact is devoured by the higher-level *Weltanschauung*. Further, the very act of creating such a world view in which the observer can observe the conflict also creates a strong conviction about its truth. It is a very common experience that is being portrayed here. The mother sees her two young sons quarreling over who should play with a toy; she changes their environment to the playground and the conflict becomes absorbed into a larger view of the world. But also the mother knows she is right to have stopped the squabble, because the very stopping of it convinces her she is right.

Now we can see the origin of the conviction in the thesis: it arose because the thesis was a larger view of some other conflict, and just because it was a larger view, it created the mood of conviction. There is nobody who feels more right than the person who can see that an argument is based on a narrow view of reality and that he holds a broader perspective. For the first time in our story, there is also a loss of seriousness and the gain of a bit of humor in this episode of inquiry. The "bigger" mind that "objectively" views the conflict runs the risk of being silly, of concocting a large but ridiculous world view. A "bigger" mind observing an international dispute may "see" that it is brought about by hidden forces from other planets, or an imperialist plot of Wall Street, or a communist plan of world domination. To be taken seriously, the bigger mind must somehow get somewhere "beyond" the opposition of convictions of the thesis and antithesis. What this "beyond" means is part of Hegel's master plan, which is an epic of the development of mind up to the stage of Absolute Mind. The bigger mind goes "beyond" the conflict when its episode fits into the larger epic. What later philosophy resents about some of Hegel's writings is the implicit assumption that he knew the epic beforehand, and thus forced the story of mind into a preconceived pattern. The mind that knows the whole epic must be the supreme objective mind we have been seeking in this chapter, and hence the designer needs to know the method by which such a mind wrote Hegel's story. Of course, Hegel himself tried to say that the epic's story was inevitable, but even so he fails to tell us how he knows this, or how he happened to come upon the correct form of the epic. In fact, Hegel fails to sweep his own mind into the story, even though his must be the most objective of all if he is right. Of course, similar remarks could be made about all the philosophers we have discussed so far; none of them is able to use his philosophy of design to account for his own mind's capability of designing.

The "bigger" mind that observes the conflict is often called the "synthesis," a term that only weakly describes the power Hegel intended to ascribe to it. Possibly the dignity we normally perceive in the role of a legal judge permits us to call this bigger mind a judging mind, and its activity "judgment." Other labels will occur to us as the subsequent argument unfolds.

Can we design judgment in the inquiring system? Again the explicit design will threaten the life of the dialectic as it goes about its task of being explicit. The next chapters will explore the explicit design question in some detail. The general idea is to design the class of models (*Weltanschauungen*) in such a way that each model can be expanded into a more general model, or else can be made more refined by introducing finer distinctions. The straight-faced inquiring system that has created a thesis and an antithesis in the manner described above now searches for an expanded *Weltanschauung* which, when conjoined with the data, makes both the thesis and the antithesis maximally irrelevant in the teleological sense. Neither is important relative to the broader objectives of the inquirer. Simultaneously, the broader and/or deeper *Weltanschauung* maximizes the credence of the "super-proposal" or synthesis. The inquirer can also work on the data bank, either expanding it or making it more precise, and search for the optimal change in the data bank that will maximize the irrelevance of the thesis and antithesis and maximize the credence in the synthesis.

As we shall see, the entire process leads to ever expanding and ever more refined models. If the search process "converges" in some sense, then the "limit" might be regarded as an objective description of reality. Why it should be so regarded is not clear from Hegel's system alone, but the idea seems to be that an approach to reality based on the most forceful arguments and counter-arguments at each stage must in the end have eliminated every conceivable ground for doubt. The world will have been examined from every possible point of view—i.e., "objectively."

Critique of the Hegelian Design

And yet there is much to make us question this design of an inquirer. We could—and will—ask why the process should lead anywhere but down blind and narrow alleys, unless there is a guide who has superior vision over the maze. The mere opposition of thesis and antithesis does not mean that the perspective of the inquirer is broad. This objection might be met in part by requiring that the expanded *Weltanschauung* of the "synthesis" be a different representation of reality in the sense of the last chapter, but even so, how do we tell whether the set of representations is free of built-in bias?

But there is a still more serious criticism of an opposite kind: Hegel's process of learning one's own mind belongs to a leisure class, where time and cost are of no concern. If in order to attain an objective viewpoint one must search all the ramifications of mind, then objectivity is a costly and time-consuming commodity; partial objectivity might be far better. Indeed, if time and cost are relevant considerations, then a mind that does not go "all the way," but instead properly balances the risks of bias against the costs, is more objective than the thoroughgoing but lavish mind.

Consider, for example, the plight of the ordinary but extraordinarily curious citizen of today. In addition to being well informed in his own business, he is called upon to vote on a plethora of issues concerning the world, the nation, the state, and the city. If he is to be a well-informed voter, he must be "fully" informed about world poverty and international politics, national economics and regional development, city traffic and educational planning. Yet one must spend a lifetime to understand any one of these topics well. The problem, then, is not, "How does the public become well informed?" but rather, "Given so much time that can be spent on any issue, what is the optimal display that can be presented to the citizen?" In the Leibnizian inquirer, the display consists of a stream of sentences (or charts), some of which may be true, others false, others irrelevant. The citizens' problem is to put together several consistent stories and then, as the data flow increases, to converge on one story that seems to hold together in the best manner.

The Lockean inquirer displays the "fundamental" data that all experts agree are accurate and relevant, and then builds a consistent story out of these. The Kantian inquirer displays the same story from different points of view, emphasizing thereby that what is put into the story by the internal mode of representation is not given from the outside. But the Hegelian inquirer, using the same data, tells two stories, one supporting the most prominent policy on one side, the other supporting the most prominent policy on the other side. The teleological issue is: Which method of telling the story will produce the optimally informed citizen when each is constrained by the same cost and time resources? The even broader issue of the well-informed public is to determine the optimal time and effort to be put into the optimal mode of displaying information to decision makers.

We are far from finding any satisfactory basis for even discussing this very general problem of teleological information. It is a problem as general as the problem of the whole system. The inclination of the thinking mind is to break the problem down into manageable parts, i.e., to classify its many components and precisely define each part. He believes that the story that unfolds can then be put together, piece by piece, into a consistent framework.

Hegel's basic theme is anti-thinking in this sense: he challenges the designer to give up the explicit. Hegelian storytelling is frustrating for the logical mind. Where does the thesis come from? It is a created episode, terribly exciting, carrying its own commitment. But the 'truth of the matter" is that the thesis is only one of a large set of alternatives that are "mapped" in some "decision-making space." No element of this space need be any more prominent than any other; how did the thesis come to be called out to play its dominating role? And what process generates the antithesis? Why the sacred number two? Surely there could be three or more competing proposals, as there are often three or more political parties. Finally, hardest to understand, is this mysterious synthesis, the master observer-of-the-subject who stalks on stage unannounced. If we could have announced him beforehand—if we could have made the conditions of his entry explicit—we could have saved all the bother of the tragicomedy of the thesis and antithesis. Indeed, once we become explicit about this master observer, the squabble between the lowlier commitments is ridiculous, frivolous, at best sadly humorous. Yet Hegel tells us the synthesis does not exist without the prior conflict: ideas are generated out of opposition.

The Storytelling Inquirer

The Hegelian inquirer is a storyteller, and Hegel's thesis is that the best inquiry is the inquiry that produces stories. The underlying life of a story is its drama, not its "accuracy." Drama has the logical characteristics of a flow of events in which each subsequent event partially contradicts what went before; there is nothing duller than a thoroughly consistent story. Drama is the interplay of the tragic and the comic; its blood is conviction, and its blood pressure is antagonism. It prohibits sterile classification. It is above all implicit; it uses the explicit only to emphasize the implicit.

But is storytelling science? Does a system designed to tell stories well also produce knowledge? Or can such a system be "designed"? Or is the storyteller ever a "system"?

We would give up entirely too much if we now gave up the explicit as a criterion of design. There is no reason as yet to declare once and for all that drama is essentially implicit, or that objective storytelling cannot be explicitly designed.

We should note in closing the most uncritical aspect of Hegel's storytelling, which flavors his whole theory of mind, namely, that the story has some point to it. The point in Hegel's case was the creation by this process of an Absolute Mind. The point is also represented in that greatest myth the nineteenth-century intellectual developed: progress. Progress is the story of mankind; men will push back step by step the domain of the unknown, ever reducing the uncertain decimal place to a certainty, ever rubbing out ignorance and superstition. Men will gradually increase the greatest good of the greatest number, eliminate poverty, drudgery, disease, unhappiness.

There must be a way to make explicit the progress that underlies the Hegelian story of mind. And, indeed, the way seems already at hand in the infinite regress of observing minds. In the words of E. A. Singer:

Suppose one were to maintain that the method of distinguishing between the "appearances" and "that which appears" was one that defined and made attainable a "real" for every "appearing," only that this "real" was no less an "appearing" pointed to a "more real" and so on *ad infinitum*. Here is no longer a circle but a progress, and if one defines the goal of this progress as an "ideal" it is none the less true that only a progress can define a *real ideal*. And it is only in the possibility of progress that one can be interested. (Singer, 1924)

Appendix

Let D be a set of "data," $d_1, d_2, \ldots d_k$.

Let W be a set of models (*Weltanschauungen*), $W_1, W_2, \ldots W_l$.

Let X be an operator conjoining an element of D with an element of W, such that for every d_i in D and every W_l in W there exists one and only one element of a set I. In other words, X maps elements of D for a given W onto a set I in a many-one correspondence (there may be several pairs that map onto the same elements of I). The set I is called the "information set" of a given W and the operator X is called the interpretative operator. Thus for each element of W there corresponds an information set, represented by I (W).

T is a set of "theses," i.e., nonanalytic sentences stating something about the world, such that no element of T implies or is implied by any element of D, W, or I.

C is a two-place function that transforms T and any I (W) into elements of the real number system. C is the "degree of credence" in T given the information contained in I (W). Hence for each I (W) there will correspond a credence measure for T: $C[T, I$ (W)]. This represents the credence of a thesis given that the world is accurately described by W.

The maximal element of W relative to thesis A of T is that of W_{oA} which maximizes the credence of A over all elements of D; i.e., $W_{oA} = \max_j [C(A, I(W_j))]$.

The antithesis, B, is an element of T which can be given maximal credence in terms of some world view and the set of data. Thus the antithesis is that T_o of T is satisfying: $T_o = \max_k |\max_j [C(T_k, I(W_j))]$ for all $T_k \neq A$.

In the case of the synthesis, we introduce an operator which "expands" each world view, i.e., maps W onto a new set W'. Similarly, we need an operator that maps D into an "expanded" data set, D', and T into an expanded set of proposals T'. The synthesis is that element of T' whose maximizing W of W' minimizes the credence in both the thesis and the antithesis.

: 8 :

THE HEGELIAN INQUIRER
ILLUSTRATED:
DIALECTICAL PLANNING

This book contains three stories to supplement its discussion of design: the organic chemist (Chapter 4), the implementer (Chapter 12), and here the dialectician. In the story of the implementer I'll have some things to say about the story writer that are also relevant here and in the earlier chapter.

This particular story is an attempt to design a dialectical experience. It was motivated by a discussion a number of us[1] had attended in Berlin, in which the theme of the well-informed public had played a prominent role. Some of those attending the meeting felt that the main problem of designing a public information system is to bring the public's informa tion up to their own in quality and quantity. We, on the other hand, were trying our best to contrast this more or less Lockean notion of information with the dialectical. Our argument was that there is a critical aspect of public policy which no one "knows." It is the *Weltanschauung* of the whole relevant system. As is usual in meetings of that sort, we made no impression on our enemies, but we impressed each other enormously, and decided we would try together to design our point of view into something specific

Several of us were working on national policies for the support of research and development. Although this problem of public policy has not been dramatized in the media to the extent that many others have, it is one of considerable importance to both the developed and developing nations. In the United States in 1966, the total federal expenditure

[1] Helmut Krauch and Horst Rittel of the Studiengruppe für Systemforschung, Heidelberg, and myself. Later on, the group was supplemented by Mrs. Marie Krauch, Barbara Kohler Peters of Heidelberg, R. M. Mason, Van Maren King, and Hilda Carmichael of Berkeley.

for research and development was estimated to be around $16 billion. But the amount is less significant than the allocation. About 90 percent of the funds were allocated to DOD (Department of Defense), NASA (National Aeronautics and Space Administration), and AEC (Atomic Energy Commission). This was during a time when the nation was becoming critically aware of its poverty problems, urban sprawl, educational burden, air and water pollution, and the spreading of revolution and dissent. It is of course incorrect to put DOD–NASA–AEC under one umbrella, but certainly the major thrust of all three is the national defense and the nation's leadership role in the world. One might imagine the nation as a person with strong leadership tendencies who was only mildly interested in his own personal weaknesses.

The policy problem is most vividly portrayed in the funding policies of specific projects. Everyone in Washington seemed to realize that if we were going to do research on how to get a man on the moon, or develop a Supersonic transport, we would have to spend billions of dollars. But if we wanted to do research on crime, sanitation, transportation, or information for a state of the union, the price tag is $100,000.[2]

Now there is a good reason (*Weltanschauung*) for being far more interested in defending oneself than in other human functions; it is the fear of an enemy who threatens to learn enough to break through our defenses. So the status quo policy (90 percent to DOD–NASA–AEC, 10 percent to the rest) depends on a world view (thesis). Indeed, it was not difficult to find a world view that satisfied the following conditions: (1) it is plausible, (2) it is dramatic, and (3) along with a set of data, it establishes the status quo policy as the best policy for the nation to follow. This is the view, expressed in a number of speeches of that time. It says that the communist nations wish to dominate the world, by military means if necessary; that no amount of negotiation will change their ambition. Hence these aggressor nations are to be treated in much the same manner as any infestation, of germ or insect, say. If there is a good possibility that the bug will become immune to our defenses, then we must pour money into research to overcome its immunity, else we'll be destroyed.

But there is also a reason why the status quo policy is seriously wrong. This reason is also (1) plausible, (2) dramatic, and (3) along with the *same* set of data, establishes an entirely different policy as the

---

[2] The amount offered for bids by the aerospace companies of California for studies of these problems in California, using the so-called systems approach. See Hoos (1969 and 1970).

best policy. Here a number of *Weltanschauungen* are available, but if we follow the speeches of the doves relative to the Vietnam war, we can piece together a world view that satisfies the criteria. It says that the critical feature of the world today is deprivation, economic and social. Not only are people deprived of food, shelter, education, etc., but through expanding communication (especially radio) they have become vividly aware of the fact that there are others who live in affluence. Also, there is no good reason why half the world should be deprived and half affluent. Obviously, the discontent arising from this irrationality creates revolution. We, the United States, being the most affluent nation, should devote our resources to overcoming the inequities; part of this effort will be research into new forms of health care, nutrition, distribution, education, further understanding of cultural differences, and so forth. Once the inequities have been reduced, then communism will lack the nourishment of discontent and will die on the vine. This second world view implies a "reversal policy," in which funds for defense research are held at a minimum to protect us while we begin to solve the economic and social problems.

It is important to note that the two policies and their associated W's are "contraries" in a logical sense: they cannot both be true, but they may both be false. There are many other policies the United States could adopt: reduced funding of research, isolationism, and so forth. But the dialectic is not an argument to establish the optimality of one among a set of plausible alternatives; rather, it is an argument which is designed to create a richer synthesis by revealing the underlying assumptions.

We designed the dialectic very carefully as follows. First a "data bank" was constructed. It contained statistics on the arms race, the race to produce engineers, Gross National Product (GNP) for several nations, military expenditures of these nations, and so forth. Since the dialectic uses mood as well as logical argument, we designed cartoons relevant to the military build-up, future urban living, and so on.

In the dialectical presentation, the first speaker (the "hawk") presents a datum (say the arms race), molds it into information with his W ("see how the USSR tries to outrun us in the arms race despite our efforts to reduce nuclear armaments through negotiation"), and thus argues for the status quo. The second speaker (the "dove") takes the same datum (arms race), but molds it into a different information through his W ("the USA frightens the communist nations by being the first to explode the atomic bomb, build the H-bomb, etc.") and thus argues for his "reversal" policy. He (the dove) then selects another datum (e.g., GNP vs. military expenditures) and shows how nations like Japan and West Germany have a higher growth rate of GNP than the United States, but a very low military budget, and this shows how economic growth is fostered by reduced military expenditures. The hawk takes the same datum, points out that Japan and West Germany would long since have been overrun by communism were it not for United States arms, and therefore ... Even the same cartoon is used to portray two different interpretations: a particularly hideous scene of homes made out of old rockets is either the future world under communism, or the world after the nuclear holocaust.

The medium for the presentation was the radio. On three successive weeks, the station presented the problem of United States' federal expenditures for research and development in three ways. The first was a straightforward discussion by a supposed expert, who had used other experts on communism, economics, nutrition, etc., to mold a sort of Lockean agreement on the best policy, much along the lines of presidential commissions. The second was a "debate" in which the debaters used different data. The third was a dialectic, designed as described above, in which both sides used the same data. Listeners were supplied ahead of time with a description of the policy issue, the data bank, and the cartoons. They were invited to send in their reactions, which included their assessment of the presentation, which presentation best revealed the underlying assumptions, and so on.

Inquiring systems with a strong bias for "unbiased" statistics will find it easy to criticize the experiment. The audience tended to be young intellectuals liberal to radical in their political attitudes (as we found from their questionnaires). The experimenters on the whole were frankly biased against the status quo policy (of course, they had to have *some* opinion). Since this experiment was a social reality, it was non-separable from other social realities. After the three presentations were made, the radio station held a half-hour "phone-in," when the audience could raise questions and make comments. Because we had announced that the research was supported in part by the National Aeronautics and Space Administration, a number of the listeners were convinced that the whole program was a plot on the part of NASA to brainwash the public. Their *Waltanschauung* says that social science has sold out to big government, and uses its so-called "objective" approach to help government interfere with individuals and their lives. We had designed a dialectic within a dialectic.

Naturally we could not completely avoid our own academic training, and were compelled to test our ideas in the more rigid environment of an "experiment." Essentially the same format was used, except that each subject was exposed to only one "treatment," and there was a "control group" which spent the half hour listening to music. Before the presentations (or music) the subjects read about the policy issue, looked at the data and the cartoons, and then wrote a one-page statement of their opinion, with brief arguments in its defense. After the presentations, they again wrote a statement of their (perhaps revised) opinions. All statements were scrambled and then "graded" by an independent group, to determine whether the after-presentation statements were statistically better than the before, as graded by the judges (who were given certain criteria like awareness of underlying assumptions). The subjects also indicated their reactions to the speakers' statements during the presentations on an approval-disapproval scale.

If one wished to play the esoteric game, he would find some fascinating and perhaps unsolvable problems in analyzing the data of such an experiment. But the major result does not require any depth analysis: the dialectical process cannot be successfully isolated from other aspects of living. If one tries to distill it out in an experiment, it becomes more or less meaningless. One might just as well remind the subject that there are hidden assumptions. One expects to create a kind of mental shock when the subject realizes that the same datum can be used to defend opposing policies; but the shock comes forcibly only if he is busily engaged with handling all kinds of information in a real social environment.

The impossibility of isolating the dialectic process was brought home forcibly in a subsequent study (Mason, 1968). A subsidiary of a large company had developed a long-range plan to meet the company's policy of return on investment and other goals. The subsidiary produced and sold a part of a larger piece of manufacturing equipment. The planners argued that the industrial market for this part was limited, as were the opportunities for an increased market share. They documented both points by a set of "facts." But the international market was far more promising, especially in certain developing countries. Hence the plan called for acquisitions and mergers in other countries. After considerable consultation with the managers, Mason saw that one could adopt a very plausible counter-*Weltanschauung,* an alternative "world of the business." For the part to work well in the total piece of equipment, considerable engineering analysis was required. In fact, many of the company's engineers claimed that the ability to sell the part rested primarily on the ability to perform this analysis. Hence, "in reality" the subsidiary was in the business of engineering services, the product being a means for selling these services. "In fact," once one perceives that the business is "really" engineering service, then the domestic market looks very promising, and the risks of international operations are avoided.

If the test of the success of this study was the amount of attention it created on the part of management, it succeeded admirably. "In fact," it could be regarded as creating a political environment, one party being the product-oriented salesmen, the other the engineers.

But it seems at least plausible to argue that the "verification" of a research project of the dialectical inquirer is *not* the establishment of a solution, but the creation of a more knowledgeable political process in which the opposing parties are more fully aware of each other's *Weltanschauungen* and the role of data in the battle for power. This argument is plausible if one accepts the world view that through the conflict of ideas comes greater enlightenment, a world view which must have its own deadly enemy, of course.

: 9 :

SINGERIAN INQUIRING SYSTEMS: PROGRESS

Metrology

The last two chapters represent a style of inquiry which its admirers would describe as soaring and to which its detractors would confer the B.S. degree. It is time for a shift in style to the more precise and explicit although, as we shall see, it is impossible to keep the vague and implicit out of the inquiring system.

The discussion of the Hegelian inquiring system ended with Hegel's optimism, the promise that the movement from thesis-antithesis to synthesis is a soaring to greater heights, to self-awareness, more completeness, betterment, progress. We now need to see if this optimism can be defended and defined.

Our resource will be E. A. Singer, Jr., and specifically his *Experience and Reflection*. Singer chose as his starting point metrology, a science which has been remarkably neglected by philosophers. Metrology is the science of measurement. Now philosophers have shown an interest in the formal language of measurement (transitivity, asymmetry, etc.), but language is only a part of the story. The really fascinating aspect of metrology from a philosophical point of view is the operational design of measurement, i.e., the steps that must be performed to produce measurements, and the justification that the produced readings accurately describe some aspect of reality.

Standards and Units

To design an inquiring system which measures, two initial decisions must be made: the unit and the standard. The unit appears to be "arbitrary," while the standard is not. As in all systems design, however, the distinc-

186

tion between arbitrary and non-arbitrary is itself a non-arbitrary strategic decision.

Suppose we use two examples to aid us in trying to design a measuring system, one physical, the other social. I want to measure the width and depth of an alcove wherein to place my desk so that I can measure my annual net income for the Internal Revenue Service. I go in search of my measuring tape (which is not where it's supposed to be, of course!), and with it in hand I compare the boundaries of the alcove with the numbered marks on the tape, and using a bit of simple arithmetic, I write down some numbers on a slip of paper. I've chosen to read these to the nearest quarter inch. Not wishing to go through the bothersome business of returning the desk to the furniture store because I miscalculated, I try two or three times with different markers, or perhaps I ask my wife to measure as well. With the desk in place, I sit and consult various records of income and expenses, using the appropriate governmental forms, and finally arrive at a net income figure expressed to the nearest dollar.

From these two homely examples, the shape of the measuring system emerges. The set of components for the length system include at least these: a rule-generating system, which specifies the steps to be followed, a tape manufacturer, a visual system capable of following the specified rules and thereby making comparisons and transforming these into numbers, and a second visual system capable of checking the first. But what is most relevant about the example is the very strong assumption that the furniture store, which presumably measured the desk for me, has very much the same system, so that their numbers and mine must agree, at least within the quarter-inch requirement. Indeed, the interesting point is that there exists a system of measuring lengths, available to anyone who can acquire a ruler or tape, which is thoroughly reliable within, say, an eighth or sixteenth of an inch. What is the design of such a measuring system?

We can readily see that the basis of the design is a Lockean community. It is interesting to note that the creation of such a community is no simple social task. In the history of the United States there was a time when an inch was not an inch or a pound a pound. It took considerable legislation, together with the formation of the Coast and Geodetic Survey and eventually the National Bureau of Standards, to bring about sufficient agreement among various sectors of the public. Even today, the numbers appearing on food packages do not necessarily represent a reliable agreement.

The key to the design of the Lockean community for measurement is the "standard." In the most general sense,[1] a standard consists of a set of operations which in principle will resolve any disagreements arising in the community. Imagine, for example, that I have purchased a five-pound bag of sugar, but on weighing it at home I find it to be only four and a half pounds. I return to the store, where the manager weighs it on his scale at five pounds. In principle, assuming a sufficient quantity of patience, we could resolve our differences, say by going to the nearest drugstore where finer weighing machines are available. But why would we believe in this method of resolving the issue? Because we might both be confident that the druggist is honest, with no stake in our quarrel, and that he is constantly checking his balance against "standard" weights. These weights themselves have been carefully prepared to conform to national "standards."

But here we seem to be on the verge of an infinite regress. Suppose, to continue the example, that the druggist decides in my favor, but the grocer, who is a man of principle even though an incredibly bad entre-preneur, wishes to check the druggist. Together we go to the National Bureau of Standards, which weighs the bag in its carefully controlled laboratory and reports a reading of 4.5238 pounds. Where does the grocer go now if he's still convinced he's right? He could, of course, go to an international body, but eventually the process must stop. Thus the Lockean community is designed so that its members agree, say, that the National Bureau is the ultimate check on any disagreements. Does this mean that the Bureau sets arbitrary units and operations? Of course not. It is the responsibility of the Bureau to assure itself that there is a sound theoretical base for certifying that a given method of measuring is, or is not, acceptable wherever it is applied, and under whatever condi-tions. This is why the "unit" of length, for example, is not arbitrary at all. One aspect of the Bureau's measure of performance is the simplicity or cost of maintaining the system, together with the degree of refinement of measurement the system produces. The shift of the standard of length from a platinum bar immersed in a liquid to the wave length of yellow cadmium was based in part on these considerations.

Here again, the emphasis in the literature on the formal aspects of

---

[1] If I were a general semanticist, I'd have to admit that the word "standard" is used throughout in at least two senses, the more general one referring to the operational design of the system, the more specific to some property of an object, e.g., a platinum bar and its markings. I hope the ambiguity will not bother anyone except a semanticist, because the context should make it clear which meaning is being employed.

measurement has led to some linguistic confusions. Formally, it is true that any unit of length can be chosen and shown to be proportional to any other unit. But it does not follow that the unit of length is "arbitrary" in the measurement system, any more than the dollar is arbitrary, if "arbitrary" means that alternative choices are equally valuable from a design point of view.

A Measure of Performance of the Measuring System

We can begin to see how a measure of performance, and hence of progress, might now be defined. Assume that there is a positive value of measuring length to a group of people, G. G includes housewives, carpenters, plumbers, manufacturers, scientists, surveyors, etc. We might then say that the measure of performance of a measuring system, M, is the degree to which M can design G into a Lockean community, i.e., the degree to which differences about length among G's members can be resolved by M.

But the lessons of the last few chapters show us that creating a Lockean community does not necessarily imply that knowledge will thereby result. Why should we suppose that the community of measurers is describing reality? A number of responses can be made to this question, as we shall see. At a very simple level, one could adopt a pragmatic position, as did John Dewey, and say that the measuring system measures reality if the use of its data "works out satisfactorily." Thus the measurement of the length of my desk accurately portrays reality if the desk fits.

It is to be noted that this account has a very peculiar twist: the measuring system is based on relatively precise rules and theories, while its defense is based on the very imprecise concept of "works out." The weakness of the philosophy is apparent enough. Most United States automobile drivers might have agreed that the internal combustion engine has "worked out satisfactorily" until they learned of its contribution to air pollution. But if one tries to go beyond Dewey to measure the real utility of length measurements, then there is another peculiar twist, for now the reality of all measurements depends on the "fundamental" measurement of utility, i.e., on a measurement process which, according to the criterion given above, has a very low measure of performance. To return to the illustration, I sit at my well-measured desk to measure my (real) income during the past year. To be sure, there are rules to be followed and observations to be made; furthermore, there will be a disinterested observer, an auditor of the Internal Revenue Service, to check my observations and obedience to the rules. But there is no Lockean community, because except in the simplest cases, few would claim that the final number "measures" income. If "income" means real value received over a period of time, then it is safe to say that no one knows how to measure income even approximately. Thus the proposed base for a satisfactory measure of length, namely, the real value of the length measuring system, is itself in a dubious state of development.

And yet, despite the fact that we cannot even approximately state the worth of our global system of measuring length, it seems absurd to say that there is a serious question about our ability to measure length. Hence, some other criterion is needed to convince us that the Lockean community of length measurers is describing reality rather than illusion. And the criterion seems to be ready at hand once we accept the wisdom of examining the history of a system in considering its design. Two hundred years ago the Lockean community could agree on a length measured within one thousandth of an inch. Today, the accuracy can be within 100 millionth of an inch. In and of itself this result is not impressive, of course, because refinement alone is hardly the hallmark of reality; today's realists scorn the scholastic ability to estimate the population of angels within one or two angelic heads. But it is worth noting how refinement does carry its own conviction provided agreements of certain kinds are possible.

Readings and Replications

To return to the bag of sugar, if the grocer and I disagree on the first decimal point (e.g., 4.9 vs. 4.5), then the druggist may settle the matter for us because his scales agree consistently to the third decimal point. In general, when two measuring systems disagree in the nth decimal point, their disagreement may be resolved by a third measuring system accurate to the $(n + 1\text{st})$ or higher level. Of course, this principle does not hold unless we have agreement in the community about certain aspects of the three systems. Our design task is to try to understand these aspects.

The key design feature of the length measuring system is the ability to "replicate," i.e., to go through the same set of operations several times. Suppose, following Singer, that we call an output of one set of operations a "reading." Then the design specification seems to say that the readings should be in "sufficient" agreement. It is reasonable to argue that if they are not in agreement, then the system is not reliably describing reality. The converse, of course, is not so obvious: if the replicated readings agree, we cannot infer that the system is working properly. To make this point clear, imagine one of the following four conditions: (1) the measured object remains the same in length over the period of time in which the replications occur, as does the measuring rod; (2) the object fluctuates in length while the measuring rod does not; (3) the object remains the same while the rod fluctuates; (4) both fluctuate. Suppose, also, that the operational rules of the design system are the simplest: compare the markings on the rod with the limits of the object and, using arithmetic, report as a reading the differences. In the first case we could assume that the readings would sufficiently agree if the observers were careful. In the second and third, we would expect trouble, because the replications would probably not produce agreements. But in the fourth case, we might find agreement again, e.g., if the object and the rod were made out of the same temperature-sensitive material in an environment where the temperature is fluctuating. It is important to notice that the four conditions are the framework of observation of another system, the Hegelian over-observer. Our question of how the system should behave in each of the four conditions is thus Hegelian in kind: how can the over-observer be created?

Apparently the simplest cases are the second and third, where the measuring system is clearly out of phase with reality. One would expect that a "competent" observer would produce "inconsistent" readings when he made "independent" observations. The descriptors "competent," "inconsistent," and "independent" are judgments of the over-observer, who judges whether the operational steps have been carried out correctly, and whether the observer's previous responses are influencing his present observations. As system designers, we might be tempted to say that two or more readings are inconsistent if they are not exactly alike. But this would be a tactical error of design, the error of naïve empiricism which tries to base all inquiry on agreement. To be sure, provided the observer is really competent and is really making independent observations, then conditions 2 and 3 cannot hold if the readings are all alike *within the level of refinement of the readings*. But an inquiring system faced with an endless set of identical readings would never be able to determine whether condition 1 or 4 holds, or whether 2 and 3 hold at a more refined level of observation. The situation is a very familiar one in all experimentation which permits replication of observation. The experimenter wishes to test a hypothesis, and finds that his readings are in agreement with his theory within a specified level of refinement. No amount of additional testing with the same results would ever enable him to decide whether another hypothesis, also compatible with the data, is false, or whether his own would fail at a higher level of refinement.

Partitioning (Refinement)

To Singer the tactical lesson seemed clear: whenever all readings are identical, then the inquiring system must shift to a higher level of refinement. It should be emphasized at this point that any such tactical decision of the inquiring system, like all tactical and strategic decisions of any system, involves an ontological commitment. In the present case the inquiring system commits itself to the idea that every meaningful descriptor of natural objects can be "partitioned." We say that a descriptor P is partitioned into descriptors P_1, P_2, . . . P_n if the following hold:

1. If X is P_1 ($i = 1, 2, . . . , n$) is judged to be true by the inquiring system, then so is X is P.
2. If X is P is judged true, then either X is P_1 or X is P_2 or . . . , X is P_n is judged true.
3. X is P_i and X is P_j ($i \neq$ j) is never judged true.
4. $n > 2$.

One interpretation of these stipulations merely says that a partitioning is an exhaustive and inclusive logical division of a set into at least two parts, but this is a special case of more general conditions. The inquiring system may use set theory as a basis of its judgments, but it need not do so. Often in the history of science the judgment has been based on a Lockean community agreement which goes beyond logic (e.g., in physics that there are exactly two kinds of particles, or in chemistry that there are n elements, or in biology m species, etc.).

The ontological assumption of partitioning is often expressed in terms of "quantification" because the number system provides a very convenient way of satisfying the four stipulations. Indeed, the essence of the "qualitative" is captured by the ontological assumption that nature can be reduced to a set of descriptors which cannot be partitioned. As we have seen, the qualitative assumption poses awkward, but not necessarily insurmountable, problems for the inquiring system. This is a point which we shall examine in the latter part of the book when we speculate about the problems of inquiring systems. Although quantification permits a very elegant way for the system to explore alternative explanations of natural events, it may also exclude a whole aspect of nature, e.g., the unique individual who cannot be pursued down the endless pathways of partitioning.

Singerian inquiring systems, then, are quantitative in the sense specified above, so that the rule to partition whenever complete agreement of readings occurs is assumed to be a meaningful rule in all cases (although it may be extremely difficult to implement). The rule is applied until the system reaches a level of refinement of its readings where not all readings agree.

Now if the readings disagree at some level, e.g., in the third decimal place, how should the inquiring system decide which of the four cases of the relation between measuring rod and object (see page 191) actually holds? The question is one of the "analysis of variation," i.e., of deciding whether a variation or disagreement is significant or not. All Singerian inquiring systems face this problem, whether the inquiry is about lengths, the planning of urban housing or computing income taxes. In the case of length measurements, the system may take advantage of the immense technology of statistical "analysis of variance," which is a special case of the analysis of variation, based on a theory of randomness of natural events. In areas like housing and income taxes, the technology becomes one of politics and law. We see a new dimension in the Lockean community, which in effect creates disagreements in order to attain a higher level of agreement.

But has the partitioning rule gained us anything? Here again the answer to this strategic question depends on a whole system judgment. In its simplest form, the assumption says that if two contrary hypotheses are both consistent with a set of adjusted readings at a specified level of refinement, then there exists some higher level where one (or both) will fail to be consistent. But this simple form is rather deceptive, since it does not take into account the tremendous resilience of general hypotheses about the natural world, nor the strong relationship between hypotheses and readings. Indeed, when the inquiring system decides that a hypothesis is not consistent with a set of readings, it may adopt one of the following policies: (1) revise the hypothesis by adding new variables, or changing the functional form of the hypothesis; (2) revise the procedure of adjusting the readings (including discarding one or more of them as being incorrectly obtained); or (3) tolerate the inconsistency until more evidence is available. Hence the role that partitioning plays is to bring the inquiring system to a stage where it must decide between these alternatives; the more sophisticated whole-system assumption says that refinement of readings will eventually produce this stage.

Kant's Problem: Design the Process of Revision

We can now appreciate the most subtle and difficult design problem of Singerian inquiring systems, which, in honor of its originator, might be called Kant's problem. It is the problem of revision of the a priori (Kant) or *Weltanschauung* (Hegel) or natural image (Singer): when and how to revise? The design problem depends on the response to the teleological question—why revise?—which in turn depends on the purpose and measure of performance of the system.

Actually, Kant's design problem goes back to the Leibnizian and Lockean inquirers as well. Leibnizian inquirers permit a kind of competition among world views, or fact nets, so that the design of when and how to revise becomes a consideration of the relative weight of each competitor. In Lockean systems, the design idea is to create a community of reasonable men, whose agreements become the basis of when and how, and even why. The community seems to work best when it does not make explicit the grounds of its agreements. But Kant and Hegel try to make the inquirer self-conscious. Kant argues that the community shares a common a priori mode of shaping and interpreting sensory responses (time, space, causality, etc.). Implicit in Kant's argument is the question whether the shape imposed on the data is appropriate. Once we pass beyond Kant's own reply (there is only one way to shape the data), we are in the land of the strategy of design with no clear guideposts. Hegel's design suggestion is just the opposite of Locke's: whenever the community builds up a strong agreement in a *Weltanschauung,* then create the counter-*Weltanschauung.* What Hegel leaves unanswered is the question whether such a procedure of disagreement gets us anywhere.

With Singer, the design problem becomes much more explicit than with any of his contemporaries. Most philosophers of science of Singer's time were devoting their energies to a "logical reconstruction" of science, using the new and very powerful tool of symbolic logic. In the language of this book, they were trying to determine how science has been designed. They were wise enough to see that science is *not* what scientists do, because scientists, being human, are often foolish and perverse even when they are "doing science." Rather, the logical reconstructionists believed that they could cull the essence of the scientific method by sorting out the inconsistencies and confusions through logical analysis. Thus they believed that there has been a basic design of science, and that the design structure can be excavated by removing all the rubble. The success of the logician in revealing the design structure of mathematics probably gave considerable reinforcement to their conviction. But the logical analysis of mathematics at best revealed only the design features of proof and not of discovery, i.e., revealed how problems ought to be solved, given the conditions, rather than what problems ought to be solved. In systems language, the logicians learned something about the tactics of mathematics but comparatively little about its strategy. In the area of empirical science, the venture was successful at the tactical level if one could assume a warranted data base, i.e., a set of "atomic" (nondecomposable) assertions about the natural world which are unassailable. Since it is almost always strategically unsound ever to design an inquirer which commits itself strongly to accepting a data base, the tactics of logical reconstructionism have very limited application. The strategic error of logical reconstructionism, for Singer, lies in its attempt to reconstruct the inquiring system by the use of only one discipline of inquiry, logic. Singer, on the other hand, saw the necessity of using the whole scope of inquiry to aid in the design task. As we shall see in the remainder of this book, the definition of "whole scope of inquiry" is itself a difficult and elusive problem, but it is almost certain that the whole scope is not limited to any one discipline, or, indeed, to all the disciplines as they are recognized today.

To pursue the underlying ideas of Singer's design, we should explore at greater length his idea of "adjusting" readings by returning to the four simple relationships between the measuring rod and the object measured. Suppose the measuring system adopts a natural image in accordance with the first type of assumption, namely, that the measuring rod and the object-to-be-measured remain invariant. But suppose, also, that the readings are judged to be significantly different. At this point, the measuring system is faced with a strategic problem, as we have noted. Suppose it chooses to change the image to option 2, that the rod remains invariant but the object changes. In doing so, the measuring system must create an image which stipulates how the object changes with time or some other measurable variable. The situation is a common one in industrial quality control; to test a lot of bullets, for example, one takes a sample, fires them through a "standard" barrel, and takes readings of the velocity. However, the object being measured (bullet velocity at the end of the barrel) changes over time or, more precisely, with the number of bullets tested; the decline in velocity can be taken as linear by the measuring system. Once the coefficients of linearity are estimated, the measuring system is in a position to estimate, for each reading, what velocity would have been obtained had that reading occurred on the first trial, when the barrel was brand new. Thus the measuring system is able to take the *i*th reading and "adjust" it back to the first reading. In other words, the measuring system has been able to adjust condition 2 (changing object) to condition 1 (invariant object) by adjusting the imagery.

At this point, those who hold precision and certainty as high values of the inquiring system may feel that the whole foundation has slipped. Once the measuring system engages in the game of adjusting imagery, and hence data, to "save" its view of the world, all fundamental control seems to be lost: there is no ultimate court of appeals. One has only to recall the very flexible and subtle strategies open to the Ptolemaic geocentric theory to see how far this game can be extended.

But such a reaction arises out of the kind of parsimony that no longer is suitable as a criterion for the design of inquiring systems. The parsimony is based on the erroneous theory that authority or authorization is essential for design. The word "authority" derives from the concept of leadership, a component of the system to which one can turn when in doubt. It is similar to the concept of control, which implies that a component can observe and correct the behavior of the system. But Singerian inquiring systems have no such component. Put otherwise, authority and control are pervasive throughout the system and have no location; the system is controlled, but no component is the controller. The idea has already been mentioned several times under the labels "tactics" and "strategies"; a tactical decision assumes an authority while a strategic decision does not. Thus a Singerian inquiring system must encompass the whole breadth of inquiry in its attempt to authorize and control its procedures.

Revision Opportunities: The "Sweeping-in" Process

Singer describes one such process, which he labels a "sweeping-in" operation. In the example cited above, where the object changes, the measurer can "sweep in" variables and their laws which enable him to adjust his readings. One sees that it would be very helpful if the inquiring system had a catalogue of opportunities in this regard, and that the traditional problem of the classification of the sciences might provide some clues. Singer's method follows a traditional one of starting with logic and noting the dimensions added by each science in turn. Thus arithmetic adds number and numerical laws; geometry adds point, line, plane, etc., and the laws of space; kinematics adds time and pure kinematical laws; mechanics adds mass and mechanical laws; physics adds groups and fields and statistical laws ("randomness"); biology adds function, organism and purpose, and teleological laws; psychology adds mind and psychic laws; sociology adds groups of minds and group laws; ethics adds ultimate purpose and moral laws.

The sweeping-in process consists of bringing concepts and variables of this catalogue into the model to overcome inconsistencies of the readings. Thus, in the examples cited above, temperature and barrel wear, both physical variables, were incorporated into the measuring system's image of nature. In the nineteenth century Bessel was able to account for discrepancies by sweeping in the reaction-time of observers, a psychological variable. We see again that Singer's design idea is one more way of building Leibnizian fact nets, and that one may view the history of the design of inquiring systems as the elaboration of the basic design features of the Leibnizian inquirer.

The construction of this catalogue of opportunities is a very difficult design task, as can be seen in the literature dealing with the topic. Some logicians dispute the contention that arithmetic "adds" anything new; relativity theory asserts that kinematics is nonseparable from geometry; in quantum mechanics, statistical laws are taken to be basic (so that mechanics and physics are nonseparable in the catalogue); molecular biology struggles with the problem of teleological and deterministic laws for biology, while computer sciences cheerfully use teleology (e.g., in problem solving) to describe the behavior of machines. Of course, a great deal of the dispute depends on what one means by "adding" a new dimension. Here Singer himself seems to be confused, because sometimes he regards the new dimension, e.g., number, to be a primitive (not definable, say, by the concepts of logic), while sometimes he regards it to be definable (e.g., he defines purpose and life in terms of physical concepts).

Nor is it clear what the progression of the sciences means from a design point of view. One might say that the inquiring system should explore as low as possible in the progression before going to a science at a "later" stage. But such a strategy would be foolish. For example, it is well known that one reason why inconsistent readings are obtained between laboratories following the same measurement procedures is the different training of the observers. It would be foolish to explore physical variables to account for the inconsistency when this more or less obvious sociopsychological variable is available. Furthermore, there is no sound reason why the inquiring system should "start" with logic. To be sure, all inquiry uses logic, but then, as we have seen, all inquiry uses every branch of inquiry. Logic itself can be regarded as a derivation of social communication, i.e., as a branch of sociology.

Sometimes the catalogue of inquiring system concepts is likened to a lattice framework of interconnected concepts, but this analogy only weakly portrays the depths of the problem. The complexity of the interconceptual design is better illustrated in that episode in physics when wave and particle imagery were recognized as legitimate dual *Weltanschauungen.* To be fanciful, the catalogue program calls for interpreting chemistry as a teleological science (so that, for example, the fragmentation of the sample in Chapter 4 is an attempt to minimize some variable of the system); or it calls for interpreting physical particles as living things; or it calls for conceiving all scientific laws as moral laws; and so on. All of the recent hue and cry for "interdisciplinary research" by foundations and other supporters of science might be regarded as a response to the collective unconscious realization that human knowledge does not come in pieces: to understand an aspect of nature is to see it through "all" the ways of imagery.

The Strategy of Agreement Revisited

We can begin to sense the endless process of the Singerian inquiring system. This feature of its design can be emphasized if we examine further the strategy of agreement. We have already seen one departure from the terminating strategy of the Lockean inquirer, when all the readings are alike. The argument was that an increasing number of like readings did not increase the system's confidence in an hypothesis, because there exist counter-hypotheses which are also in agreement with the readings. This argument extends to the case where the readings differ, but the differences are judged to be satisfactory. At such a stage, the strategic question is whether or not the system should seek a counter-hypothesis. The spirit of the Hegelian inquiring system on which Singer built his theory of inquiry says that when all is going well, and data and hypothesis are mutually compatible, then is the time to rock the boat, upset the apple cart, encourage revolution and dissent. Professors with well-established theories should encourage their students to attack them with equally plausible counter-theories. This is the only pathway to reality: whenever we are confident that we have grasped reality, then begins the new adventure to reveal our illusion and put us back again in the black forest.

But the process is dialectical, which means that two opposing processes are at work in the inquiring system. One is the process of defending the status quo, the existing "paradigm" of inquiry, with its established methods, data, and theory. The other is the process of attacking the status quo, proposing radical but forceful paradigms, questioning the quality of the status quo.

Singer in the quotation at the end of Chapter 7 called the "real" an "ideal," and we can see why. The idealist is a restless fellow who sees evil in complacency; he regards the realist as a hypocrite at times because his realism is unrealistic. The realist, on the other hand, accuses the idealist of being impractical, because his insistence on destroying the value of the present way of life precludes positive action. The Singerian inquiring system does not seek to resolve the philosophical dispute, but, on the contrary, seeks to intensify it.[2]

The Teleology of Inquiry

Singer made the theme of endless process a central one in his philosophy; his name for the restlessness he had in mind is "contentment" (Singer, 1936). What appear to be opposites, the restless and the contented, become the opposite sides of the same idea when we realize that "contentment" comes from the Latin *continere,* to "hold together." The contented life is the complete life, made up of all those aspects of a life

---

[2] I tried to portray the drama of the dispute in Chapter 14 of *Challenge to Reason* (1968).

that make it meaningful. But to be restful is to establish oneself in only one sector of a life and to ignore the rest. So to be "contented" is to be restless.

But "restless" does not really capture the essence of Singer's idea because it too often connotes pointless, whereas the Singerian inquiring system is above all teleological, a grand teleology with an ethical base. If we use the scheme on page 43, the following characteristics emerge:

1. The inquiring system has the purpose of creating knowledge, which means creating the capability of choosing the right means for one's desired ends.

2. The measure of performance is to be defined as the "level" of scientific and educational excellence of all society, a measure yet to be developed.[3]

3. The client is mankind, i.e., all human teleological beings.

4. The components have been the disciplines, but the design of inquiry along esoteric, disciplinary lines is probably wrong, as we have seen, if the purpose is "exoteric" knowledge, i.e., knowledge that goes outward to be useful for all men in all societies.

5. The environment of the inquiring system is a very critical aspect of the design. Singer's theory of value is essentially "enabling." That is, ethical values are based on an assessment of man's capability of attaining what he wants, and *not* on an assessment of the goals as such. Thus the ethical system apparently passes no judgment on the quality of a man's life. But this appearance is deceptive because one man may want to deprive another of his life or liberty. Hence the environment which the inquiring system critically needs is a cooperative environment, where A wants that goal which will aid B in attaining his goals. One sees how fuzzy the boundaries of the inquiring system become because inquiry is evidently needed to create cooperation and cooperation to create inquiry. This is why the design of a Singerian inquiring system eventually becomes the design of the whole social system.

6. The decision makers are everyone—in the ideal. But at any stage there will be the leaders and the followers. For Singer the most important decision makers are the heroes, those inspired by the heroic mood to

---

[3] Singer used to speculate on the suitability of using the standard deviation of a physical constant (e.g., the velocity of light *in vacuo*) as a surrogate measure. But this speculation was made in an era when physical science was held in high regard, and it was not naïve to expect that the findings of the scientists would be published and aid all men in the pursuit of their goals.

depart from the safe lands of the status quo. More needs to be said about these men and their moods when we assess the inquiring system vis-à-vis the concept of progress.

7 and 8. The designers are everyone—in the ideal. Progress can be measured in terms of the degree to which the client, decision maker, and designer are the same. This stipulation may seem odd in one regard, at least. If the client is all mankind, then how can those who have died be served by the living system? Worse still, since the ideal is never attained, the system must inevitably fail to serve all clients. But the thesis that once a man has died he can no longer be served is not a tautology, and indeed may be challenged by the counter-*Weltanschauung* that all men are immortal in terms of being clients. It is not even necessary to postulate individual immortality. To worship one's ancestors may simply be the act of regarding their life's intentions as being as sacred as our own and our progeny's.

9. I have purposefully stressed the theme of betterment in the foregoing account, even to the point of a kind of simplistic optimism. It is doubtful whether Singer himself would have so strongly expressed his hopes for mankind. The counter-argument is most strongly reinforced when we ask for the nature of the built-in guarantor which gives sense to the optimism.

Science and Imperatives: The "Is" and the "Ought"

The fact that the Singerian inquiring system has no real terminating point on any issue brings out some interesting features of its language. The language of such an inquiring system needs to convey both what has been learned and what has yet to be learned. In a language like English the indicative mood of expression ("This apple is green") is reasonably capable of expressing what has been learned, but is very poorly designed to express the unlearned. Singer suggested, instead, that the language of the inquiring system requires a departure from the form "X is P" as regards all three of its parts: subject, verb, and predicate. To express the uncertainties of a finding, one needs to convey the idea that the subject in the inquiring system's finding may not be the real subject which a specific question about nature has raised. The predicate should somehow express the latitude of uncertainty about the descriptor, e.g., by conveying some range of possible values. Finally, the verb should convey the information that the finding is a judgment of a Lockean community, based on its self-imposed rules.

In place of "X is P," Singer therefore suggests something like "The object observed is to be taken as having property P plus or minus ε." The "is to be taken" is a self-imposed imperative of the community. Taken in the context of the whole Singerian theory of inquiry and progress, the imperative has the status of an ethical judgment. That is, the community judges that to accept its instruction is to bring about a suitable tactic or strategy in the grand teleological scheme. The acceptance may lead to social actions outside of inquiry, or to new kinds of inquiry, or whatever. Part of the community's judgment is concerned with the appropriateness of these actions from an ethical point of view. Hence, the linguistic puzzle which bothered some empiricists—how the inquiring system can pass linguistically from "is" statements to "ought" statements—is no puzzle at all in the Singerian inquirer: the inquiring system speaks exclusively in the "ought," the "is" being only a convenient *façon de parler* when one wants to block out uncertainty in the discourse. As a computer programmer would say, the whole design is instructions, including the "data base."

Progress or Process? The Heroic Mood

Singer's theory of progress is far more subtle than the theory of "linear progress" which was popular in the nineteenth century. To understand it, one needs to adopt a dialectical point of view. On one side, call it the light side, is production-science-cooperation, the trilogy of nineteenth-century optimism. The progress toward this trilogy is toward a world of enlightenment, where men have the means to live out their individual lives in their own unique ways, without having to disrupt the lives of others, or, more strongly, with the natural urge to help others to enrich their lives. But the lessons of history tell us that when production and science begin to dominate, then society becomes fragmented; only some men reap the benefits and they do so by exploiting the environment and their fellow man.

"Oh," says the scientist, "then we must use our science to see how we can get men to cooperate more, to reduce population growth rates, air-water pollution, labor exploitation. The measure of progress must include cooperation, which cannot be separated from production-science. Refining our measures and producing more effective machines is not progress if thereby more conflict occurs. In other words, progress is not linear, but a very complicated nonlinear relationship between the enabling forces of production, science, and cooperation."

This is all very well, but one cannot help noting who is speaking: the scientist. He wants to make science, i.e., the inquiring system, the leading edge of progress because for him there can be no progress without understanding. Even if we grant him his premise that science has created more and more knowledge, why should we also grant him his other premise—that the net benefit has been positive? Why not simply say that making knowledge is like any other form of life: it happens and it is neither good nor bad. You make knowledge, he makes love; you both simply live out an existence.

To Singer, such a charge to the scientific community is based, not on so-called scientific evidence but on a "mood," a complex of emotions which arise out of man's ancestry. Had Singer written later, he could have used the wealth of material which Jung and his followers have collected to illustrate the force of the "collective unconscious" on the human psyche. Singer found his clue in but one albeit important aspect of this force, the heroic mood. Joseph Campbell has well described the structure of the mood in his *Hero with a Thousand Faces* (1956). The myths of the hero, he says, begin with some stable state of affairs, a comfortable house, beautiful wife and children, high respect, in short, plenty of production-science-cooperation. Then comes the impulse for the adventure or quest, sometimes in the form of a message from the gods or other heroes, but in any event the hero has no choice but to go forth, to leave the comforts for a kind of cold darkness. Beasts and evil spirits keep challenging him in the dark forest. In our drama, the black forest and its challengers are the mood that progress does not exist, that it is only a process at best, that the enterprise is no enterprise at all. For the hero in the midst of his journey has no assurance that anything will happen except his own death and that of his companions. At this stage the idea of progress and fulfillment seems very foolish indeed. The stage need not be tragic or ominous, of course; it may be humorous, playful, silly, lovely. Then science and its big serious program of knowledge, control of nature, and the rest look utterly ridiculous: fat science proclaiming it will save the world while it odoriferously defecates in public.[4]

But then the hero—or some heroes at least—arrive at their goal,

[4] For the contrast, see James Hillman (1968).

fight the ultimate battle, and win. As in the case of the Buddha, the battle may be a spiritual one, or for our inquirer, an intellectual one. But this is not all: the hero must return, and there is usually much to tempt him to stay and not bring back the fruits of his labors, just as Newton hid many of his important discoveries in his study. For the journey back means leaving the heights of heroism for mundane, boring, everyday existence. Furthermore, the trip back is usually another black forest and its challengers, but this time the other side of the forest is dullness.

It is very important to note that the hero's journey is not restricted to great men or to semi-gods. The hero is in every one of us, and it is impossible to say whether a Newton or Theseus is a greater hero than the individual who risks his security in the quest for self-knowledge. To be sure, the heroic mood is often suppressed by other emotions and thoughts; to free it in every man is an ideal, the ideal of a unified decision maker, client, and designer.

But what about the question: is there progress or merely process? Which is the same as the thematic question of this book: does the inquiring system generate knowledge of reality or its own form of illusion?

The response is: it depends on where you are. If you are at home, in the status quo, there is a kind of quiet progress, an orderliness, cleanness, comfort, in which little discoveries here and there push back the decimal places and provide better ways of doing things. If you are on the road, then there is no progress, just change, which can be bright or dark, funny or sad, tragic or comic. The rules are gone, laws make no sense. If you are fighting the battle, or whatever the mission may be, you are risking your soul for something overwhelmingly important and central. Progress is no longer diffuse, but here and now in your actions; revolution is one word for it. If you are on the way back, you may be disillusioned, angry, dead in spirit, or playful, or senile.

The Guarantor

Can we design the heroic mood? Jung, in his *The Undiscovered Self,* tells us about two views of the human psyche. In the one, man is counted and classified. The wonder is the diversity, but out of the diversity comes the need to lead, to pass regulations which tell us which classes of people can do what, regulations which become the State. The other world view is the unique individual and his relation to something more wondrous than himself. One might be tempted to say that design belongs only to the first view of the human being, but this would be much too hasty a judgment. The hero's quest, which is universal across mankind, is one example of the unique relationship of an individual to his god; it cannot be "designed" by any of the typical methods of design which we have discussed thus far. But design is very young, practically a baby. What would design have to be like for us to be able to design a unique individual's relationship to his god, or to design an heroic mood?

We have come by a long route back to the issues of Chapter 1, where we placed design and creativity together to examine their similarities and differences. The entire excursion could be regarded as a search for more understanding of these two dialectical concepts; the question remains the same in kind but is a book long in its asking: can design grasp the essence of the creative in each one of us?

I don't know any sensible response to this question, although I think the question itself is sensible. I could try the head-on approach of defining the illusive concepts that have crept into the Singerian design while I wasn't watching: hero, mood, tragedy, comedy, unique and god, among others. Then I'd define design, and there we'd be. At least we'd be moving, processing. But my mood suggests another kind of adventure. Very often, I've found, in the tales the hero spends an incredible amount of time just wandering around, apparently getting nowhere, or worse, being blown farther away from his quest. The approach is circumambulatory, a marvelously long word for confusion. So in the remainder of this book I'll walk around the issue of a meaning of design which could encompass the heroic mood and other aspects of the creative.

Part

: II :

Speculation on Systems Design

: 10 :

THREE BASIC MODELS
OF INQUIRING SYSTEMS

The circumambulatory excursions mentioned at the end of the last chapter begin with an old theme, the imagery inquiry uses to describe nature. The central point is easy enough to state. Science in its attempt to carve up nature and measure her pieces properly has found certain images to be acceptable—specifically, deterministic, random, and teleological imagery. The speculative question of this chapter is whether the limitation of imagery doesn't preclude our ever being able to describe the unique, the creative, and the like.

Democritean Imagery: Mechanism

Of all the many ways that men have found to look at the world that they inhabit, perhaps the most attractive to the thinking man is that first most fully developed in western science by Democritus in the fifth century B.C. The Democritean image of reality carries with it both an elegant simplicity and a real hope that no aspect of the whole real universe can escape being subsumed under its magnificent imagery.

According to Democritus, the natural world is basically composed of atoms which differ in only a very small number of properties, namely, their position, mass, and shape. Implicit in the Democritean imagery is also the concept that there exists a sufficient amount of information concerning the atoms of the world so that in principle all of their movements are predictable in the future and describable in the past. This Democritean dream of atomic imagery, as we know, had its renaissance in the sixteenth and seventeenth centuries and has carried on down to the present day. It has been the *Weltanschauung* of the physical sciences, kinematics, mechanics, nuclear physics, etc., as well as chemistry and the many branches of biology in which biological change is described in terms of predictable changes of certain elements of the organism. Very often in the history of science, the Democritean image was thought to be "deterministic," because it was supposed to provide absolutely precise predictions about events, in which no probability concepts occur.

One very attractive aspect of the Democritean imagery is its enormous capability of abstraction. By writing down in very simple form a set of mathematical equations, one can encompass vast reaches of the natural world in terms of spatial distance, spatial intensity, and time. The enormous success of Democritean imagery permits a fantastic enlargement of the basic information of the inquiring system, since a piece of information gathered at one moment of time can be extrapolated to other moments of time or other parts of space by means of the abstract underlying equations of the image.

At the present point it is futile to try precisely to define Democritean imagery or to lay out the various forms which it may take. We will see why this is futile in terms of our later discussion of the extensions of the *Weltanschauungen* of inquiring systems.

Aristotelian Imagery: Teleology

Greek times also presented to us another elegant way of viewing nature, in which the elements of nature are taken to be purposive entities, i.e., entities that seek specified goals. This is the imagery of Aristotle's *Physics*; Aristotle used the *Weltanschauung* to describe not only biological things, but the physical universe as well. In this imagery every element of nature at any moment of time is conceived as having a number of choices at its disposal, and it selects its choices so as to pursue the goals appropriate to it. In the more familiar case of biology, the living being seeks its own survival by nutrition and the construction of defense mechanisms in its environment; it also seeks the reproduction of its kind. Thus its behavior at various moments of its life can be described as attempts to gain the goals of survival and reproduction. Aristotelian teleological imagery has always been popular in the biological and social sciences. It seems to be an easy way to describe and account for the behavior of living beings. It is often taken to be the basis of psychological, mental, and social behavior, so that sciences like psychology, sociology, and economics have often used teleological imagery as a way of reporting their information and explaining the events that occur in the observable world.

Teleological imagery is also far more attractive to those who deal in the less precise forms of human living, namely, history, philosophy, and literature, in which the explanation of events and the creation of new esthetic ideas seem quite at variance with the Democritean imagery, but at least to some extent explainable in terms of the goal-seeking behavior of individuals. Finally in religion, except for a few Democritean instances of the sort that Spinoza creates, the teleological imagery holds full sway; the world as a whole is taken to have a purpose ascribable to a god, who may himself be combating evil forces with opposite purposes.

Carneadean Imagery: Probability

In the post-Aristotelian period, and chiefly in the hands of the Skeptic Carneades, there arose another kind of world imagery which has been deeply explored in the last century. Carneades adopted a basic philosophy similar to the Singerian inquiring system, where all issues, both rational and sensory, remain uncertain, but in some cases they may attain an "approvability" of sufficient strength for action. The degree of approvability becomes "probability" in modern system design.

Thus events of the world are only known with a certain degree of probability. In some technologies of system design, the degree is expressed by a probability distribution of possible states. In such an interpretation of Carneades' meaning, each state of the world can be likened to the initial state of a roulette wheel just prior to its being spun. The force of the spin and the initial state of the roulette wheel at best generate a probability distribution of events; it may happen in many cases that the probability distributions are independent of the initial states or of the characteristics of the spin, so that the possible subsequent states of nature can only be described in terms of, say, rectangular probability distributions. In this interpretation of Carneadean imagery, the world can be regarded as a gigantic gambling table in which the only laws governing the events of the world are the laws of probability.

Carneadean imagery is very fruitful in those disciplines where the welter of events and large masses of information seem to make hopeless any attempt to find a Democritean or even an Aristotelian interpretation of what has occurred. A very excellent example is the case of the stock market in western economies. A Democritean inquiring system might expect to derive from the events that occur in the stock market each day some basic set of elements that explain the changes and that could be used as a way of predicting the fluctuations in the market as a whole or in specific stocks. An Aristotelian inquiring system would expect to look at the stock market from the point of view of purposive behavior of individual buyers and sellers. It would assume that each investor has certain goals, and would try by means of its teleological imagery to predict the aggregate behavior of the market as well as the behavior of certain stocks in terms of goal-seeking imagery. Neither the Democritean nor the Aristotelian images seem sufficient at the present time to provide even the most rudimentary description or explanation of stock market behavior. On the other hand, there is real promise in the notion that the stock market behaves in very much the same way that a fairly complicated gambling device behaves; in the case of a roulette wheel, for example, it would be hopeless to search, in Democritean fashion, for the basic mechanical elements that account for the resultant end positions of the ball. Instead, the inquiring system finds it far more fruitful to look for probabilistic patterns among such events.

A close relationship has often been pointed out between the Carneadean imagery that is used to study stock market behavior and the same kind of imagery used to describe particle behavior in such physical events as, for example, Brownian motion. Here a random imagery, from the point of view of the inquiring system, seems the most convenient way of describing the events that are taking place in a gas.

There is the design question of whether empirical probability distributions in modern statistical theory really capture the essence of uncertainty; this is the same translation question which runs throughout our discussion: what is the most appropriate way to translate a historical idea, like Leibniz's monad or Carneades' "approvability," into contemporary systems design? In recent years, it has been suggested that there are other ways of measuring uncertainty, e.g., the uncertainty of an opponent's play in a rational game, or the subjective uncertainty of a decision maker (so-called "Bayesian" probability). As we shall see, it seems safe to say that none of these extensions really captures the uncertainties of the hero's adventure described in the last chapter. But for he moment, we intend that a Carneadean imagery be one in which probability is a basic concept, and probability is measurable.

Consistency and Interdependence of the Three Imageries

At first sight there seems to be a certain fundamental opposition between any two of these images of the natural world. For example, according to the Democritean imagery in its strongest form, there can exist at any moment of time but one state of nature, and this is a state of nature precisely predictable by the properties of the world at any other point of time. This characterization applies not only to the world as a whole, but to every element of the world, so that apparently any given element of nature has no set of choices of the sort Aristotelian imagery requires; nor can an element be perceived to have any goal other than in the trivial sense that from any position at any moment of time there exists one and only one pathway to its position at any other moment of time. Such a concept of goal-seeking behavior seems contrary to the spirit of the Aristotelian teleological imagery, where it is essential that a purposive entity has at least two choices. Furthermore, a deterministic Democritean imagery seems to preclude the *ultimate* need for Carneadean imagery, except again in the trivial sense in which the probability distributions are all characterized as having their weights at one point; among all the possible states of nature at any moment of time there is one that has associated with it a probability of "1" and all the others have "0's." But again this is contrary to the spirit of probability imagery, where it is essential that there be at least two states with probabilities greater than zero.

Now there is one obvious extension of Democritean imagery that will sweep in teleological and probability imagery in a perfectly successful fashion as far as the Democritean inquiring system is concerned. This extension can "explain" both teleology and probability in a fairly straightforward fashion.

There can be no denying that self-conscious living animals are often convinced that at least some of their behavior is purposeful in the sense that they do make choices. But one can explain this conviction in Democritean imagery by showing that the natural determinants of people's behavior have created in them a "choice-seeking imagination" as a way in which they respond to their environment. In other words, their own self-consciousness of their purpose is itself a predictable end-state of certain kinds of genetic origins and adaptability to their environment, and is thus predictable from the initial state of the organism and its own growth patterns.

In the same fashion the Democritean imagery can account for probability distributions in terms of a current status of inquiring systems. Wherever it is impossible for the inquiring system to dig deeply into the precise elementary nature of a mechanism, it can describe many similar types of mechanisms differing within morphological ranges. Thus, when one pulls the arm of a slot machine one may regard the particular slot machine as one of a class of slot machines with certain characteristics, and the force of the pull as belonging to a range of forces. From the description of these classes and ranges, one may generate the relative frequency of certain events that will occur when people pull the handle.

Similar extensions may be made of random imagery to develop Democritean images by suitable specification of probability distributions in which the weights associated with certain events turn out to be ones and zeros. And the extension of teleological imagery to account for Democritean or Carneadean imagery is obvious enough once one begins to realize that a particular way in which an organism tries to adapt to its environment is to try to view its environment from a deterministic or a probabilistic point of view.

As we mentioned in the last chapter, the Singerian inquiring system succeeds in far greater depth in showing how Aristotelian imagery can be regarded as a logical extension of a Democritean imagery. Singer, in *Experience and Reflection,* gives a detailed account of the derivation of the concepts of function and purpose from physical concepts. Singer's attempt is similar in kind to Rosenbluth, Bigelow, and Wiener's (1943) approach to "purposive machines," which set off the more or less futile debate about whether computers think and are conscious. The debate was futile from a design point of view, since the issue is a tactical or strategic one with a fairly straightforward answer in many cases: can one use and develop computer capability better by employing teleological concepts rather than mechanical concepts? Singer also explains a way in which probability imagery can occur within a Democritean image.

In the physical sciences, the relationship between Democritean and probability imagery has been deeply explored. In quantum mechanics, the physical scientist found he needed probability imagery to account for certain observable events in particle behavior. A number of technical problems then arose concerning the relationship between the various types of imagery that had classically been available to the physical scientist and the kinds of requirements imposed by the use of probability imagery.

Up until recent times there was little interest in the scientific community in extending teleological imagery to incorporate the Democritean and probability imagery. As was indicated above, the manner in which this can be done is to explain Democritean imagery as a choice of an organism in its attempt to grapple with its environment. In this case, then, concepts that are taken to be ultimately basic in Democritean imagery, such as particle, wave, time, or space, would be defined within a teleological image of the natural world. Even logic itself can be regarded as an extension of a teleological image in which one starts with the natural world of teleologically communicating entities.

We now see why it is futile to attempt precise definitions of any of the three images described in this chapter. In a sense any one of the world images can be regarded as an "extension" of the other image, and all three images are in effect part of the strategy of the inquiring system. A given imagery strategy not only describes the world, but also determines the way in which information about the world is stored and retrieved by the inquiring system. If the designer of the inquiring system is Hegelian, he must sweep into the imagery of the inquiring system the image of the system itself. As was indicated in the last chapter, a certain kind of contradiction seems to have occurred historically in this regard because inquiring systems that are particularly prone to adopt Democritean imagery still regarded themselves as purposive entities in which creativity (in this instance, unpredictable behavior) is regarded to be an essential aspect of themselves. With our increasing success in extending Democritean imagery to incorporate teleological imagery and vice versa, this discordance in past designs of inquiring systems will begin to disappear.

It might appear as though the future of the design of inquiring systems is a bright one indeed. The major disputes of the past concerning whether nature is "basically" deterministic, or "basically" free, or simply the roulette wheel of an indifferent god, all seem to fade into insignificance once we begin to realize that the various images are the choice of the inquiring system. We realize that wherever it becomes difficult for the inquiring system to account for its information, it has at its disposal an enormous capacity of extension, where extension simply means a modification of the basic *Weltanschauung* in accordance with certain rules.

Limitations of the Imageries: The Fourth Box?

We have not shown, however, that an inquiring system faced with an unsatisfactory state of its information necessarily has available to it a pathway which will remove the unsatisfactory state. In other words, we have not shown that, given the enormous abilities of inquiring systems with respect to extensions of their *Weltanschauungen,* they inevitably have the ability to create new solutions to old mysteries. They *do* have the ability to extend the number of possibilities open to them, but this does not promise that the inquiring system will inevitably find the solution to all of its problems.

Put otherwise, as we saw in the last chapter, we have not been able to show that the inquiring system can understand its own creative process as opposed to its own "ritual" process. Perhaps the three basic models described in this chapter are only a subclass of a much larger set of models available to the inquiring system. By what pathways could the inquiring system determine and construct new kinds of model building? Perhaps, after all, in our story so far the inquiring system has only come to learn the most unimportant part of itself—namely, its manner of collecting, adjusting, storing, and retrieving information in accordance with the three imageries and their extensions. It may even succeed in accounting for the origin of its data and the role it itself plays in the construction of data banks and their characteristics. But the larger mystery remains: How can the inquiring system guarantee its ability to cope with all its enormously difficult problems; where is its guarantor?

There is still another mystery about itself that the inquiring system has so far failed to face. The inquiring system, it is true, tries to understand the world in a knowledgeable way. But there is a world which is not an inquiring system. Some of this world is made up of people who have quite different purposes from that of understanding the world in a knowledgeable way. Perhaps—and just perhaps—the rest of the world does not want inquiring systems. Could the inquiring system determine whether its own activities were desirable to the rest of the world, and if they are not, could the inquiring system determine why they are not?

In a way, this question seems to be most inappropriate for an inquiring system to pose to itself. The great traditions that have come down to us from the fifteenth and sixteenth centuries have always incorporated the notion that knowledge is good in and of itself. To know some aspect of nature that was hitherto unknown, to explain something that had hitherto been unexplained, are goods that need no defense. Consequently, when an inquiring system asks itself the question whether its own activities have any value in the total world system, there seems to be a direct and simple answer: pure knowledge is purely good.

And yet in recent times this answer has been attacked on several counts. From a purely economic point of view, the political system, which can now be taken to incorporate the inquiring system as a part, will view inquiry as only one of its many resources. As large-scale inquiring systems become expensive, the political-economic system will tend to examine the value of the activity and compare it with the value of other types of activity. But suppose the inquiring system, which is a clever fellow, tries the following tack: "All right," it says, "let's compare the utility of research and development with the utilities of other activities of the total social system. To do so, we obviously need an inquiring system." Thus the inquiring system wants to convince the total political system that inquiry is a way of determining whether inquiry is worthwhile. But such an argument for a well-designed political system is always suspect.

After all, the public has come to realize that the impact of applied science on our environment has not been completely beneficial, to say the least. It has often run roughshod over the human spirit and its culture. Science and technology have supplied the human being with many techniques to satisfy his wants. They have also supplied him with a much uglier, much more frustrating world to live in. The world appears to be a far more dangerous place for the human being to inhabit because of technology. On simple biological terms, the human animal might feel inclined to regard the inquiring system as a cancerous growth which is in need of being removed by means of a serious operation.

The Strategic Choices of Global Inquiring Systems

These charges against inquiring systems suggest several choices. One, of course, is to abolish them—an intolerable and indefensible choice for this book in these times. Another is to argue that the basic concept of design is suitable, but the restriction of the imagery to the three reviewed here is wrong. A third, which we shall explore for the moment, is to turn the inquiring system from the esoteric to the exoteric. Esoteric inquiring systems find their criteria of choice inside the system, e.g., in the statistical inconsistencies between the readings discussed in the last chapter. Exoteric inquiring systems use the criteria of non-inquiring systems to establish the satisfactory or the unsatisfactory nature of the inquiring system.

Thus, from the exoteric point of view, the design of an inquiring system requires that it be examined from the much larger perspective of industrial production or public interest or ethical standards. In a way this design idea is another manner of describing Descartes' fundamental worry about the ultimate validity of an inquiring system's information. But now, instead of suspecting that the inputs to the inquiring system are the result of the malicious behavior of the devil, one now begins to suspect that the whole notion of information itself, in social terms, is a notion that implies a basic evil for the society rather than a good.

These comments therefore suggest an extension of the inquiring system so that it can view itself as a part of a much larger system and, in particular, view its role in terms of the behavior of the larger system. This extension of the inquiring system is often called "implementation."

: 11 :

IMPLEMENTING
INQUIRING SYSTEMS

Science and Its Code of Conduct

The scientific communities of the western world since the Renaissance have struggled to formulate their own code of conduct. Above all, the scientist is expected to be honest in the sense that he reports what he has observed as truthfully as possible, and he draws his conclusions in accordance with an explicit logic that can be checked by his colleagues. No chicanery is to be permitted in the scientific community. Over and above this, the scientific community, itself and alone, stands as a judge of the quality of a man's work, especially in the pure sciences. It judges whether or not the work is of sufficient importance to be published; if it is published, it judges the kinds of rewards that should be extended to the researcher, based on its own esoteric criteria of relevance, elegance, simplicity, and generality.

As a consequence, the scientific community has long fought off any attempts by the rest of society to enter its arena and to judge its inhabitants according to external social and moral criteria. If a religious community, for example, believes that a certain kind of scientific investigation is dangerous from a religious point of view, this belief is taken to be irrelevant evidence by the scientific community. And if the politician tries to interpret the results of scientific inquiry to his own political advantage, the scientific community regards the interpretation as biased and at best useless, at worst dangerous.

Nonetheless, there are points where even pure science seems strongly to impinge upon the rest of society. These points occur when the investigations of the researcher disturb the biological or social scene in some manner unsatisfactory to the rest of society. This seems to be well illustrated in cases where social scientists attempt to investigate the attitudes and opinions of other members of a community. It is also true where biologists by their experimenting threaten to disturb the organic life of the natural world. In these cases, it is not so much a question whether the evidence is valid as whether the act of collecting the evidence disturbs some other aspect of society, e.g., individual privacy or the natural environment.

In recent years it has become more and more obvious to some social philosophers like Jacques Ellul (1964) that the activities of physical science have vastly changed the whole economic and political structure of our society, so that our lives are determined by the applications of the discoveries made in the last two centuries within the physical sciences. Some of these discoveries have a very dangerous tinge to them, of course; many a common man today may look upon the discovery of nuclear energy in the last few decades as amounting to nothing but evil for the human race.

How concerned should the scientist be with what the social community from time to time takes to be the implications of his scientific work? Should he pay due regard to the challenges which are made by his society, and should he supplement his esoteric code of conduct with exoteric criteria?

These are questions. Since they are questions, to any inquiring system theme must occur the additional question whether problems of this sort are susceptible to inquiry, and if they are, whether the inquiring system should give them high priority. Is the question of science's moral responsibility to society more important than the question of the nature of space or the genetic code?

Science and the Political Community

It is very difficult to discern exactly how modern science has responded to these challenges. Certainly many scientists believe, as did their scientific forebears, that the internal moral code of the scientific community is sufficient to sustain the activities of the community, and that the few minor exceptions where scientific activity seems to impinge upon other activities of the social community should be handled more or less on an *ad hoc* basis. These scientists are often appalled by what they call "big science," which is perfectly willing to accept the largesse of the political community and consequently bear all of the criticisms and investigations that the political community may from time to time feel necessary in order to justify its large expenditures for research and development. The history of the United States space program is a case in point. The "purer" scientists have often questioned the scientific importance of the program as it was designed. In recent years the decline of funding has brought about an economic depression in the aerospace industry, with the result that many scientists and engineers have become jobless.

Many purists in the scientific community therefore feel that science should return to the "one-man concept," in which the scientist develops, more or less on his own initiative, the particular kind of inquiry he wishes to conduct and eschews large funds and a "team" approach to a large-scale research project. The purist regards science still to be the output of one creative mind; he thinks that the most compelling motivation is curiosity, and the most satisfactory mode of judgment is elegance and simplicity.

Scientific Advisory Panels

Nevertheless, big science is certainly here, and there can be no question that the total social community has a great deal of interest in its activities. For example, many developing countries in United Nations discussions are highly suspicious of the word "research" because to them it means a diversion of funds to the esoteric community. As a consequence, those scientists who believe that we have entered into an age where large expenditures are required in order to advance science must recognize the need to understand and respond appropriately to the political community. They do this in a rather odd way: they serve on various kinds of United Nations, federal, state, and local "science policy" committees, where they give advice to legislators and government agencies concerning the funding and utilization of science and engineering. They try to judge the particular role that a given governmental organization *should* have with respect to the scientific community. This advice is frankly given as a matter of opinion, and a few of the leading scientists who have entered into this kind of activity would claim that their advice-giving is based on the same kind of careful inquiry that they conduct in their own scientific domains. The more conscientious among them may insist that there be a staff function which undertakes investigations of one kind or another, but even these staff inquiries are conducted with a quite different degree of control from that which they would deem appropriate in their own work.

We have already seen how difficult the information problem is relative to policy making. Specifically, the data by themselves never dictate a policy; very strong assumptions must be made about the whole system. But is the scientist in a particularly favorable position to make these assumptions? He may begin to feel that he is if he gains some extensive experience in the political community.

Thus the situation is odd because the scientist has had to turn into a type of politician. He must now be quite careful what he says in the panels and other advisory committees on which he serves. He must also recognize himself as in some sense defending a particular policy of allocation of federal and state funds to certain kinds of research. If he is reflective, he will see that he is a lobbyist for a particular type of scientific investigation. He becomes overconcerned with "rocking the boat." Or, if he is a dominant type, he takes upon himself the role of delivering a number of lectures on the importance of certain types of scientific investigation. He may even go so far as to "put the spot on" certain of his colleagues whose research and other interests he does not personally condone. In some cases he has been willing to taint members of the scientific community with being soft on communism, with the hope that therefore the ambitions of his scientific colleague-enemies will be thwarted. In any event, whether he plays a mild or a fierce politics, the scientist finds himself in a situation that calls for a totally different type of moral code from the one traditionally found within the pure sciences.

To all this confusion about the role of the scientist is added the often unexpressed assumption that someone who by dint of his brilliance has risen to the heights of prestige in science is thereby better qualified to make policy judgments than his mediocre colleagues. This assumption is almost obviously false: brilliance alone is never the basis for sound policies. Sound policies depend more on ethical and moral acuity than they do on bright ideas alone.

Whither Science and Policy?

But perhaps we have been witnessing a more or less temporary state of affairs, a response of the scientific community to vast and relatively rapid changes in the political scene. As we have seen in the study of Singerian inquiring systems, if for one reason or another the scientist has found himself incapable of understanding a certain segment of nature in a satisfactory manner, he has sought to extend the knowledge of the inquiring system by following one of the many pathways open to him. Perhaps, then, in the present situation where there has been an increasing need for science to respond to actions in the political arena, the scientific community will begin to recognize the high priority of understanding societal change (see Platt, 1969).

Of course, understanding science and its relation to society is an old matter for science, since even in the earliest days of modern western science the scientist was often asked to explain certain parts of nature so the entrepreneur or the politician could thereby create either the means to manufacture goods or the means to destroy his enemy by applying the findings of science. A great deal of this applied science obviously had a beneficial effect on society, e.g., the great discoveries about man's anatomy and the subsequent discoveries of bacteria.

What might be called the traditional approach of applied science, however, has always been one in which the applied scientist tries to select a given segment of the total social community and investigate how improvement of a certain type can take place in that segment. Because he has felt himself to be an extension of the scientific community, he has usually been called upon to take a rather modest attitude with respect to total, real improvement. Thus the doctor advises the patient to go home to bed, but does not feel called upon to make the ultimate policy decision and feels no particular responsibility if the patient fails to comply. The doctor merely feels that "all other things being equal," a rest is called for, but the patient must himself decide whether all other things are indeed equal. The engineering profession applies scientific discoveries to the construction of buildings, roads, bridges, etc. Here again an attitude is adopted of "all other things being equal." One particular type of bridge is to be taken as better than another. The engineer does not typically extend his advice-giving to deciding whether or not the bridge itself is appropriate or is appropriately located in the place that the city planners have decided; the city planners themselves do not generally take on the enormous task of deciding whether the allocation of funds to traffic expansion is more appropriate than the allocation of funds, say, to expansion of health services by an urban community.

Hence the need to understand the sociopolitical community has often been regarded as a very restricted need, not requiring much more than common sense and a good ear. However, even in this narrower concept of science, it has been no easy matter to develop suitable moral codes relative to the scientist's behavior in society. The codes tend to be quite general. They call upon engineers, psychologists, doctors, etc., in addition to maintaining the objectivity of scientific investigation, to take "due regard" to possible harms that may occur if certain applications are made. Generally the codes do not concern themselves with the appropriate mode of investigation to determine whether or not harmful effects will occur in the community as a whole.

The Counter-Weltanschauung: The Perils of Science

There is at least a plausible view of the world of inquiring systems which says that the politics of advising politicians and managers that have been followed in the past are seriously inadequate. According to this world view, there is a growing danger that inquiring systems may be in serious trouble with respect to the social community. The basic problem is the design-non-separability of the components of the total social system; apparent improvements in one segment of the system often lead to situations that are more troublesome, or even irreversible in their damage, at a later point in time. Our ability to create nuclear energy obviously vastly enhanced our ability to "control nature"; but it also weakened our ability to "control nature," namely the nature of international competition. Another example is the enormous advances that have been made in prolonging human life, advances certainly of importance to many an individual; but as soon as the prolongation of human life occurs without any concurrent improvement in economic or environmental conditions, the so-called "improvement" amounts to increasing the length of an impoverished and suffering life.

The *Weltanschauung* says that modern science has now become the servant of the politician. Since so much of politics is a dehumanized, anticultural kind of nationalism, science has come to be a political conspirator in a vast social degradation, the puerile selfishness of the international squabble.

It may be true, then, that inquiring systems are capable of developing highly refined esoteric methods of evaluating their own output in terms of elegance and generality. But the rest of society often comes to regard the inquiring systems as being either socially evil or, at best, a useless and highly subsidized esoteric activity of a segment of society.

It's as though a group of artists set up their own criteria for the beauty of their works and succeeded in spreading their paintings across the landscape, thereby polluting the environment.

The Contrasting Viewpoints of Science and Society

In other words, the esoteric scientist says, "This is true, or as near to true as our esoteric methods permit." The rest of society says, "It may be true in your esoteric sense, but since any exoteric application of it is either useless or harmless, it is false in our sense."

To the esoteric scientist, the viewpoint of the rest of society is either wrong or irrelevant. The "fact" is that nuclear fission takes place under such-and-such conditions; no layman from the rest of society should even presume to know enough to refute this fact because he has no knowledge of physics. But from the exoteric point of view, the so-called "fact" becomes a weapon, a threat, a power of some kind; if the weapon, threat, or power is dangerous, then so is the "fact," which is therefore "false." A politician could say that no physicist should even presume to know enough to validate the fact because he has no knowledge of politics.

Suppose, now, the physicist replies, "We must make a careful distinction between the validation of a fact by empirical means and the use of that fact by society." The astute politician's answer is: "Why must we?" Indeed, there are some very compelling design reasons why we should *not* make any such distinction because the two processes, validation and use, are non-separable in the total system design. To make a distinction or not to make a distinction is itself a policy decision. In a way, the scientist tends to be unaware of the power of his materials, i.e., ideas; "merely" to distinguish between ideas is to do something which is not mere at all. Remember, a "fact" is a communal entity; to make it communal is to bring it into the community, which may have all sorts of repercussions.

Inquiry into Exoteric Criteria:
System and Behavioral Science

From the design point of view, suppose we accept the dialectic of the challenges that external society makes of science, and assume that developing exoteric criteria is a top priority of inquiry. Now in some sense there seems to be no difficulty in this regard whatsoever. In the last chapter we have pointed out how the inquiring system is capable of extending its models and particularly how extensions into the domain of teleological behavior appear perfectly feasible. Indeed, in recent decades just such extensions have occurred in a number of different types of inquiring systems—in operations research, behavioral science, systems science, and so on. In these cases the inquiring system takes upon itself the task of determining: (1) the goals that a segment of society holds most valuable; (2) the alternative means at the disposal of the decision makers; finally (3) it recommends the particular plan of action which it concludes is most satisfactory to the segment of society that is relevant.

This activity of the inquiring system goes far beyond the policies of applied science mentioned above in connection with engineering and medicine because now the inquiring system tries to determine what the goals of the particular segment of society really are, and it does not recommend in the spirit of "all other things being equal." Indeed, if it fails to take as large a viewpoint as possible in terms of its capabilities, it can be criticized by other scientists because of its failure adequately to state the problem. Thus a freeway designer who merely uses the criterion of speed of flow through a highway artery can be severely criticized by his colleagues for his failure to recognize the impact that a particular design of a freeway may have upon the surrounding communities or the dangerous features that may be introduced. The new systems designers, the operations researchers and management scientists, try to attain as full an understanding of the decision maker's problem as possible. This they do by extending existing models of decision making in greater and greater breadth and depth in order to encompass varieties of goals, alternatives, and environmental constraints.

It is to be noted that this effort of the operations researcher may be an implicit criticism of the other policies of science mentioned above, of advisory panels and limited-scope applied science. In a sense, the operations researcher might accuse his fellow scientists of selling their services cheaply so that the customer gets a product that won't work.

In the case of the enterpreneur in western society, these efforts of the systems designers may turn out to be remarkably successful. Here the goals can often be subsumed under the generalized heading of "net profit," or a suitable extension of this concept, like cost-benefit analysis, across a long period of time. Other goals of the society in which the entrepreneur exists can be represented by various limits on the entrepreneur's choices, so that he is constrained, for example, by certain legal forms, by considerations of safety, and so on. Subject to these constraints, the systems designer attempts to develop a plan of action which will maximize the profits of the entrepreneur over a period of time. Efforts of this sort have already brought about a revolution in manufacturing policies and to some extent in the marketing and financial policies of corporations. A somewhat similar but differently organized effort has taken place on the part of certain behavioral scientists who have attempted to study the ways in which men create decisions together and to suggest organizational changes and changes in personal relationships which will create a more effectively operating organization in terms of the organization's objectives.

In all of these instances the scientist undertakes a task of far greater magnitude than any that applied science has ever tried before. There seems to be no limit to the size of the system to be studied in attempting to develop a plan of action. Nor can the scientist in this instance any longer regard himself as being detached from the decisions that are made. Even if he himself does not make the decision but only makes a recommendation, he assumes as heavy a responsibility as the decision maker does, once he makes his recommendation in the form of a particular plan which he considers to be best.

The management scientist often discovers that what he considered to be a sufficiently closed system for the development of an optimal plan turns out to have facets and modifications that he had never suspected. He finds, for example, that people identify with their roles in a certain manner, so that when new courses of action are suggested, their most typical reaction is to look at their particular role and the way in which it will be changed if the new plans are adopted. Top management may believe that the total organization is striving to maximize some function of money. But the individuals in the organization may also attempt to maximize a particular kind of relationship they hold to the organization. The latter may very well be threatened by a plan which increases the total organization's effectiveness. These people, therefore, resist any suggestions for change. Since they are often in a better position than the scientist himself to conduct internal politics within the organization, they can often stifle the proposed plans for change by various internal organizational mechanisms.

Although the models of science discussed in the last chapter can be extended to include teleological behavior of the sort just described, it is doubtful whether any of the current extensions reach in sufficient depth to understand the motivations of the people who live in organizations, and as a consequence many of the existing models that have been developed for "optimization" simply fail to provide the basis for improvement. It is clear that chances for acceptance of a recommended course of action depend in part on the way the inhabitants of organizations perceive the recommendation. If they perceive that the recommended change will leave the organizational structure and the roles within it more or less invariant, then there may be relatively little opposition to a change and little political struggle to thwart it. For example, suppose an operations researcher develops a linear programming method for the scheduling of a production plant. But the recommended plan of action, elaborate though it may be, and to some extent mystifying in the eyes of many of the people in the organization, may not be resisted simply because the members of the organization do not perceive that the recommended course of action in any way threatens their own position. They may even be somewhat awed by the elegance of the underlying mathematics and consequently pleased with the notion that they might support a change in the organization based on such a marvelous mathematical elegance of reasoning.

But this turns out to be a rare story. Changes that are not perceived to imply organizational changes are usually of a fairly trivial nature and amount to small improvements in terms of the goals of the organization. Far more difficult problems occur when the suggested changes in the organization involve the mysterious unconscious and conscious goals of various persons and groups of persons within the organization. It is simply the case that we at the present time have no real insight into the adequate design of an implementing inquiring system, that is, a system capable of carrying on its inquiry so that the implementation of its (exoteric) recommended conclusion will occur with maximal success.

Who Learns, Society or Science?

The feeling tone of the last paragraphs was intentional. It says that science has much to contribute, and needs to extend its mode of inquiry into certain aspects of the social domain. In the next chapter there is a story of one such attempt. The story is told as though this were an experiment to determine what psychological and social factors prevent the implementation of science's findings. It's the same question that people raise when they ask why there is so little "technological transfer," why perfectly sensible innovations don't become adopted. The idea seems to be that if we can find out what prevents society from learning what's good for it, we can work on these blockages. We (the scientists) can educate them!

But you can also read the story another way. The subjects in these experiments were also rather astute inquiring systems who were trying to learn about themselves and the experimenters. The counter-*Weltanschauung* says that the subjects have a better model than the experimenters, perhaps a model that is not an extension of teleology and probability. If so, it is the experimenters who have benefitted from the transfer. The idea of convincing the public to accept the Supersonic Transport may be twisted around the wrong way; the public knows what the engineer and his political allies have yet to perceive: the SST doesn't fit into the future.

In view of the ever-expanding scope that the planner is required to consider, and of his inability to apply existing designs of inquiry to handling the enormous problems of scope, it is only natural for him to think in terms of a kind of faith about the future. He will want to believe that his failure to consider everything relevant will not end in disaster. He will want to believe that his plan is better than no plan, and is the best that men can find in the present circumstance. He will want to believe that most men can be trusted to have sufficient regard for each other so as not to undertake the wanton destruction of each other.

In fact, he will need, this planner, a kind of faith in nature, a hope for his plan, and a charity with regard to his fellow men. He will need to incorporate a religion into his inquiring system. Can he do this without fundamentally changing the design of the system? Can religion be modeled by one of the existing *Weltanschauungen* of inquiring systems?

: 12 :
IMPLEMENTATION: AN EXPERIENCE
IN EDUCATION

There is nothing contained in the meaning of experience that is not already to be found in the meaning of experiment.

—HENRY BRADFORD SMITH

Two Educational Theories

One way to look at the design of inquiring systems is to regard the theory of their design to be a theory of education. One traditional theory of education is modeled along the lines of a Leibnizian-Lockean inquiring system: the members of the established community initiate novices into the community and develop their fact nets. Consciously, at least, the process is one-directional in the sense that the teacher instructs the pupil to agree to the agreements of the community. Of course, "instruction" does not mean a forceful imposition, but rather a guidance to the pupil who of his own free will decides that the teacher is right. Another theory of education asserts that the process should be two-directional: both parties are teacher and student. Furthermore, there is no terminal point of the process (e.g., agreement), but rather an enlargement of understanding, or a revision of the image, or any of the other changes discussed in the design of Singerian inquiring systems.

The Plot

We can begin the story with the first theory of education. The situation is designed as follows: a group of five persons is given a task to perform (essentially, a problem to solve) where the right performance is not easy to see. The designers of the situation (the "experimenters"[1]) form a Lockean community in the sense that they agree on the solution on the basis of their expert opinion. They select one member from the group and teach him the "solution," so that he, too, agrees. The rest of the group do not know that there is a "stooge" in their midst who has been taught by the experimenters; they regard him to be like themselves, just one of the group struggling to perform the task. The educational process which the experimenters wish to observe is the manner in which the stooge teaches the organization the solution *while the organization itself is busily at work*. This process is significantly different from the process where a teacher stands in front of a class; the class supposedly recognizes that the teacher is an expert and that the student is there to learn from the teacher. In this situation, on the other hand, the other members of the group do not recognize that the stooge is an expert, at least at the outset, and most emphatically they do not see that they are there to learn from him. Rather, they are there to perform a task.

This situation was designed to simulate an expert, e.g., an operations researcher, in an organization where the expert believes he has a solution to a critical problem and sets about trying to "implement" it. Typically, the organization does not take it for granted that the expert is right; furthermore, the members are busily engaged in performing their functions and do not think of their roles as being students of the expert.

In most of the experiments, the specific design was as follows: Each member of the group receives a set of instructions which tell him that he is an officer of a firm which makes three products, labeled P, F, and B (pencil, fountain pen, and ball-point pen). It functions by assembling these items from purchased raw materials and by setting a price for each product. Cost and time information is supplied (e.g., labor costs, production rates, inventory carrying cost, cost of raw materials, etc.). Furthermore, there is a record of prices and sales of each product for the past twelve periods. There is no advertising and one can infer from the past "sales vs. price" that there is no competition, because if sales are plotted against price, the relationship is clearly linear.

The information given in the instructions is complete in the sense that, using certain very plausible assumptions (e.g., that the sales-price relation will remain linear), one can derive the optimal decisions for purchasing raw materials, scheduling production, and setting prices so as to maximize profits. The information and the assumptions are reliable

---

[1] Philburn Ratoosh, myself, and many graduate students. For one "complete" experimental run see Huysmans (1970).

in the sense that if the subjects follow the optimal they will indeed maximize profits.

Like the mathematician in the chapter on Kantian inquiring systems, one could regard the group's task to be one of maximizing a polynomial subject to certain constraints. However, it is safe to say that very few subjects ever represented their task in this manner.

Independent Variables

In most of the experiments the subjects communicate by written message and do not see each other; hence the strategy of the stooge can be programmed, and therefore modified in various ways, while the basic structure remains the same. The experiments themselves were a part of another educational process. Over several years, graduate students in an operations research seminar at Berkeley designed different strategies for the stooge, as well as modifications of the basic design, and ran the experiments. The educational aim was to provide the student with some insight into the human side of operations research: how busy people react to technical recommendations.

The following is a partial list of the designs the graduate students invented:

1. Organizational structure (e.g., centralized [one decision maker] vs. decentralized [one decides price, one production schedule, one raw material purchasing, one investment policy, and the fifth is a coordinator]).
2. Aggressiveness of the stooge (a domineering vs. accommodating attitude).
3. Friendliness of the stooge (measured by having the subjects meet before the experiment and discuss a social problem—actually the research and development problem of Chapter 7—and then judge which of the five they liked most; subjects met face-to-face on occasion).
4. A "follow-the-leader" environment (with the idea that if some other group is visible that is using the optimal policy, the subjects will follow them. There are three companies, X, Y, and Z; X has the subjects, Y is a dummy company which simulates the behavior of other groups in the past, and Z is the "optimal" company. There is an industrial journal which publishes articles on "how we set prices" or "how we schedule production," to

which the subjects contribute. The Y and Z articles are simulations, Z's containing the optimal solution, Y's being vague and general).

5. An information system geared directly to the model (for example, the cost the group incurs in a period for holding inventory, for setups, etc.; since the typical accounting calculation of net profit is irrelevant, it is not reported).

6. A labor union (to put pressure on management to make more profit; subjects meet with the union members).

Results

Traditional reports of experimental findings are designed like detective stories; one waits until the end to see what the story is all about. But in this case it's very hard to identify the hero and the villain, since the results of the experiment were as much about the designers as they were about the subjects.

One objective of the designers was to determine whether a given "independent variable" facilitated the "implementation" of the solution i.e., improved the educational process. Now in Chapter 1 it was argued that "knowledge" implies the ability to pursue a goal even when the situation changes. The experiments were a direct application of this concept. We distinguished between "acceptance" and "knowledge" (or "understanding") of the solution. The solution was accepted if the subjects followed its prescriptions up to the close of the experiments. If they did follow them, then some slack time was created on the assembly line. If the subjects asked, they were told that they could make another product Q (for quill pen), and were given the same kind of basic data they already had for the other products. But now the stooge excused himself, and the subjects had to try to adjust the method they had been given to product Q; if they did so, they were judged by the designers to have understood the solution. (It was amazing how often the subjects, who were business school students for the most part, adopted a "cost savings" attitude; instead of asking for a way to utilize available facilities and reduce overhead on each item, they thought of ways to cut labor costs by reducing the number of shifts.)

What Was Learned About the Subjects

If one takes an overall count of the experiments and compares the net profits of "control groups" without stooges (normally run in each case) with the net profits of groups with stooges, the control groups tend to do better. That is, in the overall picture, the presence of the "man-of-knowledge" did not help and might have been a hindrance. However, this result may be deceptive, as are all experimental results; some of the experiments (like the aggressive stooge) may have produced a situation where the stooge was an active irritant.

Acceptance did occur occasionally, but was tenuous, the subjects often reverting to their own style in later periods. Understanding also occurred, but rarely. In the order of the independent variables given above, the results ran as follows:

1. Only one organizational form, the decentralized, could be maintained; the subjects could not tolerate the authoritarian form and revolted against the leader. When we tried the "team" approach where each voted on all decisions, the subjects spent the first two periods losing money fast while they debated on how to organize themselves.

2. The aggressive stooge was effectively cut out of the communication channel; the results were inconclusive for the stooge who pretended that he was merely following through on the other fellow's excellent idea.

3. The "friendly" stooges were able to gain a degree of understanding, the unfriendly were not, a result that needs a lot more analysis before it says anything meaningful. "Friendly" might mean "respected for one's opinion," in which case the result is not surprising. At one point in this experience, when I thought that no one could understand the solution, I personally handed it to one group made up of my own students and told them it was a good idea; they implemented it. (Was I "friendly"?)

4. The group did not follow the leader at all; instead they seem to have regarded Z as out of their class. But they took Y, the simulated "normal" group, to be their chief rivals, and tried to do better than Y.

5. This one frustrated the subjects greatly. They could not go on without a profit-and-loss statement; they "knew" from their training that every manager needs such a thing. The operations research-oriented data meant very little.

6. The labor union partially demoralized management: no significant change in implementation.

What Was Learned About the Experimenters

One of my students described the experiments as the "wild, wild West." The experimenters couldn't make up their minds whether they did or did not believe in traditional designs of inquiry. Thus they replicated, used control groups, tried to measure implementation, and so on. But the tasks and the environment were much too complicated to know whether everything was the "same" in a replication. Indeed, if the variance between replications is small in experiments of this type, it is just as uncertain what the inquiring system's policy should be as it is when the variance is large. If we take an Hegelian overview of the whole scene, in which we observe experimenters observing subjects, the results are quite striking. The experimenters are trying to impose a very strong legal structure on a "micro community" (Cowan, 1965). The laws of this community are really quite fantastic. Not only does the law tell you where to sit, how to communicate, what to read, but it also tells you what your (temporary) goals must be (in our case, to maximize the firm's profit). Furthermore, you are not permitted to ask why the law should hold, nor why the goals are important. If the variance is small between replications, what does the experimenter look like, according to the overviewer? Why, he must look like someone whose central purpose is to set up an environment in which people comply with a set of rules, regardless of race, sex, background, and so on. Now if the situation were a game these people liked to play, then the experimenter would approximate his ideal. So the most plausible way to represent the business of the experimenter is that he is concocting games. If the variance is small, he is succeeding in getting people to play his game; if it is large, then his rules don't work well.

The other feature the overviewer sees is that the experimenter wants very much to have other people in his Lockean community recognize that his game is "like reality." He's like a chess fan who wants people to see that chess is like war, so that taking a pawn is like wiping out a battalion, more or less. Mostly less. The overviewer is apt to let his sense of humor take over: decision-making experiments are a joke, in our case a practical joke played by the stooge.[2]

To the experimenter, who like anyone else can also become an overviewer, this reaction to his experiments may seem unkind and rather

[2] See "The Humor of Science" in Churchman (1968).

irrelevant, if not irreverent. He sees enough similarity between his game and some real-life situation to make a bold inference to reality. The argument is a familiar one: what must the whole social system be like for a decision-making experiment to be informative about how decisions are made or ought to be made? We note another familiar theme as well. The experimenter must isolate some aspect of a total decision-making system, or else the experiment is unmanageable. He therefore makes the same old assumption of design separability we have discussed so often. He must be a metaphysician if he wants to be self-conscious.

Personally, I found both these and the dialectical experiments of Chapter 8 to be rewarding and frustrating. After some time, I came to regard them as an educational technique for subjects, students, and teachers (not really distinguishable!). It is a technique which is very promising, terribly difficult, almost impossible to evaluate. In fact, it's just like any other plausible technique of education.

: 13 :

THE RELIGION OF
INQUIRING SYSTEMS

The Need for Religion in the Inquiring System

One theme that emerges from the discussion of the last chapters is that an exoteric, or implementing, inquiring system has no safe and assured pathway ahead. It lacks any objective evidence that its proposals are sound or that its method of measuring its performance is a real guide to its future. But the inference to be drawn is not necessarily a pessimistic one. There is much to be said for a faith in our future and a hope that inquiring will serve man well. Faith and hope are concepts of religion, the next place to visit on our circumambulatory walk.

But it must seem to those who reflect on the formal structure of science and its methods that bringing together the concepts of religion and of science is a reactionary movement. This is largely because the inquiring systems that underlie our religious life in the western world have their own highly developed systems of inquiry, which in certain basic respects are very much at variance with the whole spirit of science. In part, the difference is due to the particular role that so-called dogma plays in religious inquiring systems. Superficially one might feel that in the religious inquiring systems, certain dogmas are stored in the central generalizer of the system and can never be displaced or modified, whereas in the so-called scientific inquiring systems, any generalization is subject to modification or even rejection as evidence produces doubt in the "dogma."

If we go a little deeper, however, it is perfectly clear that science has its own dogmas, which maintain themselves in an astonishingly strong way despite the accumulation of various kinds of arguments and evidence against them, whereas religion, on the other hand, does modify its so-called dogmas in the light of various kinds of evidence, even the evidence collected by science itself. Examples of both trends are the struggle to maintain the dogma of the ether in physics in the early part of this century, and the changes in procedural dogma of the Catholic Church in the last two decades.

Obstacles to a Religion of Science

It is therefore probably a deeper insight into the nature of the barriers that separate science and religion to recognize that religion makes use of a *Weltanschauung* which since the days of the rationalist has ceased to have significance in the scientific community. In Chapter 10 we discussed the three basic *Weltanschauungen* that are of chief importance to the scientific community today, the Democritean, Aristotelian, and Carneadean, and their so-called "extensions." The religious *Weltanschauung,* on the other hand, describes a certain kind of relationship—such as love, adoration, and obedience—between men and other men, or between men and some superior being, or between men and "Nature." It is not clear to the scientist how any extension of teleological models could possibly translate the meaning of these relationships. Of course, a scientist like William James in his *Varieties of Religious Experience* can describe how people experience their religious life, and he can do so by an appropriate teleological model. But here the scientist is the over-observer telling us how observed people behave. The scientist, as a scientist, has great difficulty in seeing himself as a religious inquirer; he sees no clear way of subsuming science under a religious representation.

Every now and then, it is true, the scientist "steps over" into the religious area, at least in the imagery that he uses to explain his own theory. An excellent example of this "step-over" occurred with the advent of game theory in the late 1940's and its application to statistical theory of testing hypotheses. Game theory originally considered the relationship between a set of players with individual motivations and their interaction within the context of rigid rules of behavior governing the pay-offs to the individual players. It was only natural for the statisticians to adapt the notions of game theory to the testing of statistical hypotheses by conceiving of "Nature" as a player in a game against the experimenter. The "religious problem" is to determine what characteristics of benevolence or malevolence to assign to "Nature" in the game between the scientist and his great opponent. Of course this imagery may have been merely a convenient way of describing the strategies used by the statistician in drawing conclusions from a set of data. But nevertheless the justification of the strategy seemed to depend on conceiving of some kind of a Being who stands outside of the inquiring system and exercises control over its inputs. This is a crude religion of science, to be sure, but nevertheless it has some of the characteristics which we wish to discuss in this chapter.

The Science of Religion and the Religion of Science

On the other hand, there have naturally been attempts from the religious side to "step over" and to found a "science of religion." People like the humanists who have been impressed by the great successes of science in the modern age have often been persuaded that the methodology of science can be applied to areas other than the typical ones where its success has been most marked. They believe that the so-called objectivity of scientific method can be extended to religious issues. However, none of these attempts to found a "science of religion" has been particularly interesting to the scientist himself.

From the viewpoint of the present position of science, however, it seems to be far more important for the design of inquiring systems to discuss the religion of science rather than the science of religion. By analogy the same remark could also be made of the development of a "science of management," which is an attempt to extend the methods of science to the problem of managing; of far greater significance, however, as this book attempts to point out, is the development of an adequate basis for the management of science, i.e., the design of inquiring systems.

In the religion of science are included the ideas developed in Chapters 2 and 3, namely, the guarantor of the inquiring system. Specifically, in the case of the Leibnizian inquiring system, the religious part dealt with the proof of the existence of God, who then guaranteed that the so-called "fact nets" would inevitably approach unique solutions. In Leibnizian inquiring systems it was held that the designer must be able to apply the procedures of the inquiring system to the verification of the processes by which the religious component of the inquiring system acts. Thus Leibniz thought he could extend the imagery of the inquiring system to encompass religion, but our discussion led us to reject his extension as a valid design.

We should note that the designer of an inquiring system might simply reject questions of ultimate guarantees as being in some sense "meaningless" from the point of view of his concept of information and question-answering. But such a designer does not escape the religious issue. He must consider the question of the survival of the inquiring system. If he regards the question to be meaningless, then he must have a theory to substantiate his claim. His skepticism or agnosticism becomes his religious base.

Faith

Whenever I am so lucky as to get a scientist to talk about the survival of inquiry, I find he typically speaks of his faith in the enterprise. Leibniz's "proof" of the guarantor has been replaced in modern times by the much vaguer, but perhaps much more deeply felt "faith."

If we look at faith from the design point of view, we ask whether a faithful inquiring system is better than a faithless one. And here a teleological answer may be at hand. Suppose we simply say that the inquiring system may be designed to believe or not to believe any statement it examines; since this is a choice, the designer needs to know which choice is better in terms of the system's goals. The designer can argue that an inquiring system with faith in its future must be far better than one that looks upon the world of the future in terms of its own destruction. So despondent an inquiring system could do no better than to close its doors and commit an appropriate type of suicide. The decision is analogous to that of a gambler playing at very heavy odds. Suppose he is permitted to gamble his fortune in favor of a very uncertain event against the possibility that if he loses, he loses his life. There can be no point whatsoever in betting for one's own destruction, no matter what the odds. No matter how slight the chance, the gambler must in some sense have faith in the one possibility that is favorable to him. The argument is reminiscent of Pascal's argument for faith in the doctrines of the church. His point was that if the church were right and one chose not to adopt its faith, then one spent eternity in limbo or hell, whereas if the church were wrong, and yet one adopted its faith, one's soul either did not survive or its fate was determined in any case. As a consequence, one should gamble on the only possibility for gain.

This argument has very much the flavor of a formal logician's mind at work, and one can readily suspect its very basis. Indeed, its conviction simply disappears once one realizes that if the world has no guarantor, there are really many things that the human being may appropriately do that he would not do in a world in which there is a guarantor. Suicide is only one of the possible paths open to the human being in a guaranteeless world; pure play, rampant sadism, free love, are only a few of the choices. Indeed, the gambler representation of faith is based on a teleological theory of long-run goals; it is irrelevant if long-run goals are senseless and immediate existence is the highest value.

No, it must be possible to develop the religion of inquiring systems far beyond the very simple-minded type of faith described in terms of the philosophy of gambling. Faith based on gambling theory is psychologically so superficial as not to merit a mature consideration on the part of the human inquiring system. It is at best the faith of a *puer eternus,* the playful boy whose spirit often finds its way into the activities of the scientific community. The serious designer of inquiring systems, on the other hand, must look for a deeper basis for the religion of his system. It is a problem demanding a self-reexamination on the part of the inquiring system. If we were to use psychological terms, we might well say that the problem of the religion of science is a problem of the inner life of science.

Now we have already been spending many pages in the development of this inner life of the inquiring system. The pathway that stretches from Locke's concept of information to Singer's is one of great significance in our discussion of the matter of the faith and religion of science.

Inquiry Revisited

Hence, it will be wise to return once more to the concept of inquiry to see what the excursions through various designs have taught us, and thus to understand how faith might fit into our experience. Implicit in Leibniz's notion of inquiry is the development of clarity; the inquiring system struggles to find a connection between the various perceptions it has stored in such a way that the entire picture of its perceptions becomes clearer and clearer. What gives the Leibnizian inquirer its particular characteristic of inquiry is the concept of God, because God plays not only the role of the guarantor of the convergence of the nets of contingent truths, but He also stands for the ultimate objective of all of the activities of the inquiring system. Only God perceives the world with perfect clarity. If we add to this consideration the modern systems idea that all components of the system are strongly non-separable, then no creature less than God perceives anything with perfect clarity, not even the proof of God's existence. Indeed, we might plausibly argue that in a strongly nonseparable world, all contingent facts are equally uncertain: the existence of God is as certain or uncertain as the correct means to bring about world peace.

Thus if we apply some contemporary systems analysis to rationalism, it is apparent that it would be a contradiction in terms to expect the human being to have attained an absolute proof of the existence of God or of His characteristics. Whatever proof the inquiring system may attain concerning the existence of a supreme being, this proof itself must be subject to the eternal doubting that characterizes the rational inquirer. Thus in Leibnizian inquirers faith is needed to supply the gap between man's ability to perceive and perfect clarity. In the case of the supreme being, proof is errorless; in the case of man, it is always capable of error. What man requires in order to communicate with his God is some kind of faith that will compensate for man's fundamental inability to attain perfect proof.

In Lockean inquiring systems the relation of inquiry and faith is different. Since a strong Lockean system accepts the simple sensory inputs as data that are never to be modified, then there can be no faith associated with the sense data themselves, nor does a pure Lockean system seem to require any religion. It may be a matter of complete indifference to it whether or not the sensory inputs represent reality in any sense, or whether its continuous struggle to accumulate more and more information will prevail. It simply goes on creating pictures of the world as it senses the world, apparently without regard to its own destiny or even to the ultimate validity of its own findings; it may even believe that questions of ultimate validity are meaningless.

It is also true that in the Kantian inquirer, God does not enter into the empirical program of man's attempts to understand the phenomenal world. In the latter part of the first *Critique* Kant attempts to show that theological concepts, as well as the concept of faith, are not essential a priori foundations of an empirical science. They only become fundamental and essential when the inquiring system turns to questions of its own will and of the nature of the world in-and-of-itself. These questions, for Kant, lie beyond empirical inquiry.

Therefore, in the case of the pure Lockean or Kantian inquiring systems, matters of religion and faith clearly belong to the system we call the "whole man," but they appear to be of no concern to that part of man which deals with his empirical inquiry. The appearance, however, becomes an illusion once the inquirer sees that his "pure" inquiry rests on an agreement among members of a community. We argued that empirical inquiry becomes meaningless unless agreement obtains, and that the existence of agreement is not a simple input. Indeed, the Lockean and Kantian inquirers require a faith in the existence of agreement, since they cannot empirically establish its existence.

We have already seen the strong role that faith plays in the Hegelian and Singerian inquiring systems, which add self-consciousness to the design. We can conclude, then, that faith is an integral part of all inquiry, just as deception is. Faith and deception are two sides of the same process.

The Many Sides of Faith

But does this conclusion say very much? It does if the designer takes its message seriously because it says that the wise designer who uses history as a guide will perceive that the history of faith and religion is an enormously important resource for the design of inquiring systems. We have already seen how the imagery of the hero myths can be used to enlighten the mysterious mood of the individual who in the midst of plenty and satisfaction undertakes the perilous journey.

Three other examples should suffice to illustrate the power of religious imagery. In the first pages of this essay, the idea of the creative act was introduced, and we asked ourselves whether it could be "designed." We can now recall the story of Poincaré, who had for a long while been trying to solve a difficult mathematical problem. Suddenly, when he was in the midst of the very mundane act of stepping on a bus, the whole solution flashed in his mind; as Spinoza would say, he not only knew the solution, he knew he knew it. But what was the flash of insight? It's not very helpful to label it "intuition," nor to say that it was the result of "unconscious thinking." It seems far more revealing to say not that the solution flashed in his conscious mind, but rather that God spoke to him. Why? Because once the designer gets the idea that the creative thought is the voice of God, then he looks to the past to find the many ways in which other designers have tried to listen and speak to God: prayer, meditation, love, and so on. This "fourth-box" imagery is not teleological in the classical sense, though the designer's exploration may eventually enable him to relate teleology to the religious experience of communicating with God.

Have enough patience to listen to the other examples. Many scientists these days extol the virtue of basic research which has no apparent purpose other than to reveal some aspect of nature. In the eras of cost-cutting, they urge the funders to support them because who knows what fruits basic research may eventually produce; they go on to mutter about transistors, atomic energy, polio vaccines, and biochemical warfare. What a crass way to defend the glory of basic research! It's much as though a devout Catholic defended his devotion on the grounds that he doesn't want to go to hell. Why not say, instead, as our forefathers did, that basic research is one of the finest ways we have of glorifying God? A deep and significant finding about the origin of rock formations is as glorious a way to worship God as a mass, or sacrifice, or prayer, or drama. The designer may begin to see ways of appreciating basic research which were never open to him before.

This is the third example. At the end of his essay on astrology in the *Encyclopedia Britannica* (1970 edition), Benjamin Farrington tells us that: "as a serious and systematic world view claiming the allegiance of many of the best intellects in every rank of society[1] astrology is dead." Farrington wrote his piece at a time when Jung and others were finding in astrological imagery a rich gold mine of insights into the nature of the human psyche and its relationship to nature. Astrology rides again!

The Counter Thought

But all this is becoming much too direct for a circumambulatory walk. Science has spent many centuries in building its edifices and establishing traffic controls to keep the rabble out: the fanatics, opportunists, crackpot researchers, and just plain nuts. Even the Jungian heroic adventure looks perilous in this regard. Although Jung himself kept extolling the virtues of empirically sound concepts, nevertheless the imagery he allowed into his thinking constantly runs the danger of meaningless proliferation. Once we say that an "archetype" may play a significant role in our unconscious life, and that one should look into the myths and other religious materials of the past to "find" the archetypes, then there seems to be no end to the game. There is the archetype of the Great Mother,

---

[1] Our old friend, the Lockean community!

the Senex, the Puer, the hero, the beautiful maiden and her beast, the clown, Narcissus, Hyacinthus, . . . Even Jung's "collective unconscious" is at best suggestive, especially since no one to date has been able to capture the meaning of either personal or collective consciousness. Singer suggested that we return to the original meaning, "being knowledgeable with," which is the essence of Hegel's mind-observing-mind. But the formal definition seems far from capturing the experience of individual consciousness.

Should the designer run the risk of allowing charlatanism to get into the meshes of the system in order to gain deeper insight into creativity and basic research? It's safe to say that only the heroes will dare submit papers to *Behavioral Science* which use models of man-communicating-with-God, or alchemical imagery. And since the reviewers of the hero's article interpret their role to be the fight against vagueness and charlatanism, his article may never be communicated.[2]

All this is but another example of the eternity of the dialectic, in this case the eternal contrast of conservative and radical (which is conservative, which radical depends on the designer of the dialectic, of course). The dialectics of idealism-realism, mind-matter, determinism-freedom are never resolved, but they may take on new forms or even richer forms if one accepts the Hegelian concept of progress. Indeed, Hegel's own favorite dialectic, the specific vs. the universal, is a fundamental one in this essay: to understand how to design a system one must understand the whole system; on the contrary, to understand how to design a system, one must understand the unique individual.

Our teleologist, brave man, would like to step into the fight over the appropriate models of the inquiring system. He'd like to assess the costs and benefits, the assurances and risks, of extending inquiry's imagery to myths and religious experience. Thereby, he thinks, the debate can be "resolved."

Another Reflection

Thus, he thinks the synthesis of the dialectic is teleological: balance the risk (cost) of religious imagery vs. the benefit of historical insight where cost and benefit are measured in terms of the aims of the inquiring sys-

[2] But Ian Mitroff (1971) has been trying to bridge the gap between science and mythology.

tem. But why should the inquiring system have any "aims" at all? Who told the designer that it must be teleological? Teleology is a mixture of good and bad, benefit and cost, assurance and risk. Why shouldn't the inquiring system be pure, a non-decomposable entity? Or, if love is the greatest of the three, greater than faith and hope, what of love and the inquiring system? Love and purpose are antithetical, are they not? Love is antiteleological, surely?

: 14 :

PURE INQUIRING SYSTEMS: ANTITELEOLOGY

The Boundaries of the Inquiring System Revisited

In the earlier pages of this book we raised a question which we have repeatedly considered ever since—namely, the maximum size of an inquiring system from the designer's point of view. In considering the question, the designer must take into account various aspects of the world which might influence the performance of the inquiring system; if he feels that some of these aspects of the relevant world could themselves be subject to redesign, then he is forced to regard such aspects as "parts" of the total inquiring system. Thus the politics of a given government might very well become a part of the inquiring system if: (1) the politics generates the allocation of funds for research; (2) the designer of the inquiring system feels that the political systems of the country can be changed by his design.

Nor does the so-called basic scientist escape this implication of systems design. He may vastly prefer to confine himself to his laboratory and the kinds of activity for which he is most excellently trained. Nevertheless, if outside of the laboratory doors a political activity threatens the survival of his particular intellectual interests, then perhaps he may be forced to regard these external political activities as more of a challenge to his own work than a colleague's criticism of one of his theses. He cannot think that his own work ends at the end of his own career because the pursuit of knowledge is endless.

The designer of inquiring systems is therefore obliged to consider what activities he should undertake which guarantee that the work that he is conducting can be continued in another generation. The impact of the "imperative" of the Singerian inquiring system in this instance is an

247

imperative given to us by the coming generations. They demand that today's scientist pay attention not only to his own specialized work, but to the whole social and political environment in which this work occurs, so that they may also engage in similar enterprises.

Not only have the inquiring systems today become larger in the sense that the designer must consider many more aspects of the social and natural world than he ever has before, but the interrelationship between the parts seems to become more and more intricate and significant. With the advent of world nationalism, the various parts of the world feel a strong and reasonable incentive to declare their rights, to defend their peoples, and to extend the domain of their own particular interests. Within each of the nations there are threats of various sorts. There are those who fear the onslaught of communism in the West or of capitalism in the East; they try to thwart any type of activity in which they see the least sign of a swing toward what they regard to be an undesirable political philosophy. They do this partly in terms of their own interests, partly in terms of their own kind of information which, like all information, is converted into the particular framework in which they view the world. Usually these political interests are narrow in the design sense because they wish to correct one aspect of the social world, e.g., the tendency toward communism or a loss of rights within the political state.

As a consequence, the inquiring system must inevitably become closely related to various aspects of society, and these aspects are to be considered as parts of the inquiring system from the design point of view. It was this condition which brought about the speculations concerning implementing inquiring systems, and led us in the last chapter to discuss the religious life of the inquiring system.

As we noted, the problem of system boundaries arises because of the teleological orientation of the designer. The designer wishes to create a system which will function in accordance with certain standards that will eventually lead to improvement. The teleological model demands that the designer consider those aspects of the world that will influence "success" or "failure," and further demands that he consider to what degree he himself can change these aspects.

Antiteleology: The Counter-Weltanschauung of the Designer

Such at least is the concept of design which has so far guided us in this book. But, as we pointed out in Chapter 7, if the inquiring system is Hegelian, it must have the ability to look at itself, and this ability implies that the inquiring system can in some sense construct the opposite of its own image. That is to say, it has the capability of asking itself what an inquiring system would be like which did not function according to the basic principles of its own design. This is the way the system looks at itself. Otherwise the basic principles of its design become tautologies or meaningless aspects of its own life. Now the opposite of the kinds of design that we have been discussing in this book would be an inquiring system that did not consider its performance in terms of goal-seeking behavior. It would find its life in its own activities and not in what they imply.

The philosophy underlying such inquiring systems is not a new one. It has come down to us through the ages in various forms and in various kinds of expression. It is the philosophy that underlies a man's feeling of direct satisfaction in performing a leisure-time activity such as archery or golf, where the goal is not to behave so as to win; rather, the activity is good in itself. It is also the activity of the artist painting a picture, the musician composing a sonata, or the scientist engaged in a type of discovery about nature which fascinates him. It is also the mood of the religious man in his contemplation of the infinite, or of his own inner self, or of the moral man who is honest because honesty is a final good.

In this philosophy it is meaningless to ask what a particular activity implies in terms of "more ultimate" goals. Activity is a value in and of itself, and what occurs in the activity is all there is to be said about the value of the activity. If another man appreciates the paintings of someone else, then the act of appreciation, whatever it may be, has its own inherent value and has no relevance whatsoever to the value of the painting. One does not count up the instances of pleasure gained by an artistic creation or a technological development. Whatever value these may have occurs when the act of their creation takes place.

Teleological Rebuttal Number One

As we have repeatedly seen, there is no way to restrain teleology. Many people these days tell us that participation in community life is a good in itself; but as soon as the teleologist senses the popularity of participation, he immediately thinks, "Aha, *that's* how we get people to move together toward the community goal!"

So the teleologist will interpret antiteleological philosophy in his own terms. Why should a man undertake to paint, compose sonatas, conduct experiments, or whatever? The answer must be, according to the teleological model most commonly used by western man, that these activities bring a certain fundamental pleasure, or perchance they are essential for the survival of the individual in some psychological or biological sense. Their meaning, therefore, is again to be found within a teleological *Weltanschauung* of the type that we have been discussing. The teleologist goes on to point out that, whatever the artist may say about his own motivations, when we observe his behavior within a teleological model, we can explain why he sacrifices food and family for a more important goal of his life.

Nor are the other models embarrassed by the antiteleological argument. The determinist will "explain" the need for self-expression as an outcome of biological determinants, as the advocate of randomness may explain it as a biological aberrant.

Now the antiteleologist points out that "pleasure" or "benefit" cannot possibly be the goal of the artist or scientist because so much of what he does is not pleasurable or beneficial to him and in fact often constitutes the deepest of psychological distress. It is a kind of torment that the creator would never want to share nor ever want his beloved ones to experience. It is indeed a kind of inner experience that goes beyond the expression of words of any spoken language.

The teleologist, of course, is not yet defeated. He has no difficulty in pointing out that much of human experience involves enormous pain of the sort described by the creative artist or pure scientist. The point is not, he says, whether pain occurs but what would happen if the creative individual were deprived of his activity. What must happen in such a case is a kind of mental suicide, which is far worse than the torment that may occur during the creative act.

Similarly, the determinist will seek to find the reasons why certain organisms display apparently non-natural behavior; he may observe that certain animals commit suicide, or inflict pain on themselves, and he will seek to find the underlying mechanisms that explain such behavior. In an image that incorporates randomness, these deviations may be explained in terms of the random generation of genetic codes.

To all of this, the believer in the essential value of activity will simply reply that none of these images captures the true meaning of life in the immediate and unique sense. He challenges the designer of systems who is tied down to the three images described in Chapter 10, the mechanistic (Democritean), the random (Carneadean), and the teleological (Aristotelian). He claims to be talking about another conception not encompassed by any extension of these three models.

The Spirit of Antiteleology

What evidence could there be for the need for another *Weltanschauung* to add to the three that have so successfully played their roles in the development of the inquiring systems to date? Why, simply the evidence that accumulates in the writings and expressions of the human race, in its poetry, its music, its religion, and its philosophy. To all three image makers the philosopher will say, "No matter how far you have traveled down the road of your model building in order to explain my particular attitude of mind, you have come no way at all toward understanding what *I* mean by existence and what *I* mean by the good and evil of human living." This man, this antiteleologist, antimechanist, antirandomist, simply states that the deepest feelings of the human race have nothing to do with goals, or mechanisms, or randomness.

The mood we are exploring finds expression in so many ways in the human life that the reflective mind finds it difficult to state what is being talked about in terms that would be satisfactory both to the rational model builders and to the people who feel the mood in the deepest way. It almost seems as though our languages were incapable of bridging this difference between two aspects of the human psyche.

The mood of course is most familiar to western man when he reads the pages of oriental philosophy and tries to understand the utterances of the poets and philosophers. The mood has also found its way into the western world in various forms of both contemporary and ancient philosophy; it is no stranger to the western mind, although its appearance in western thought has often been quite different from its expression in oriental philosophy.

We can see that the debate is the old one between feeling and thinking. Feeling is outraged that thinking claims to have captured the essence of inner feeling by its deterministic, probabilistic, or teleological models.

The Twist of the Knife: Ateleology Is Basic

But thinking plays strange tricks on itself, without the need of feeling to bring about its troubles. Suppose we see how a bit of thought can be used to show that all three of the models require a purposeless state as the ultimate base of their theories. In this part of our circumambulation, we'll substitute the thinker's term "ateleology" for the more dialectical "antiteleology"; antiteleology is a challenge to thinking as displayed by the three *Weltanschauungen,* while ateleology is a way of thinking about them.

Suppose we represent the world in accordance with strict deterministic laws; then the events of nature are never the free decisions of any man. In the earlier Stoics, and later in Spinoza, the only possible ethics in such a determined world was the prescription for each man to understand the underlying rationality of the determined world, i.e., the explanation of the events that occur in his environment. But even this prescription is not a prescription to select, from among a set of alternatives, that one which leads to rational understanding. To be sure, Epictetus, that beautifully superficial thinker, gave both master and slave the "freedom" to adopt an attitude. But a thoroughly deterministic psychology would not even permit this much freedom in the events of nature. Instead, the philosophy of determinism would have to say that a man may appreciate understanding when it occurs to him for its own sake the moment that it occurs. According to a deterministic model, he could not possibly hope to "choose" a plan that leads to understanding. Hence in this case the extension of the deterministic philosophy to ethics simply describes a man appreciating the moments of rationality as they occur to him, in an ateleological manner.

In the same manner in a probabilist *Weltanschauung* the events that occur are the accidental conglomeration of many minute events in nature. How does it happen that I sit here writing this book at this moment of time? How many different little events must have occurred to bring about the one that is now occurring? And who would dare to have predicted two years ago that I would be sitting here as I am now? No one could conceivably have "planned" the writing of such a book.

The "futurists" who tell us about the year 2000 are a ridiculous crowd for the believer in a world of chance. There are too many conglomerations of accidents to make a specific future state have anything but a very low probability. Planning and teleology are mythical behaviors, fantasies of the minds of animals caught in the world of chance. If one of the animals appreciates a happening, that's all there is to say about it.

Perhaps the cruelest blow of all is that the ateleological thesis is the foundation of the teleological *Weltanschauung*. From Aristotle we learn that the highest form of activity of man is contemplation because contemplation is that particular function which distinguishes man from the rest of living beings; later philosophers and biologists contribute to this theme in terms of higher and lower forms of life, the concept of learning, psychological development, social improvement. But when the living being has passed from the lower forms to the highest form, then what? While he sits contemplating in Aristotelian fashion, shall he ask himself what the purpose of all this contemplation is? Such a question is meaningless except in terms of the immediately circular answer that the "purpose" of contemplation is contemplation, this being the "highest form" of his living activity.

One is reminded of Lenin's paradox: When the state withers away, and the dictatorship of the proletariat exists, then what? Wasn't it lucky that Stalin prevented such an embarrassing event from occurring?

What shall we say when a man has learned all there is to learn about a given situation? Shall he then ask himself what the point of all the learning was? Such an answer would be meaningless in terms of a pure inquiring system. To have found the answer to a question is to have answered it once and for all, and there is no point that lies beyond. The attitude of the pure scientist is that he seeks simply to satisfy his curiosity, and once his curiosity is satisfied, that is that. Curiosity is not satisfied for the purpose of creating some units of "utility." Curiosity must be satisfied, and when it *is* satisfied, there is no more ultimate goal to be attained.

So when a man has attained his aspirations, then that is that; as far as that particular pursuit is concerned, no more is to be said. His pursuits may now turn elsewhere, but the attainment of the first pursuit is not to be regarded as a means in the attainment of other pursuits.

Of course, Singerian inquiring systems avoid the ateleological terminus of teleological behavior because in Singerian design neither satisfaction nor dissatisfaction are to be taken as end states. Rather they are signs of the need for additional planning and striving. In a sense, man struggles *not* to find solutions but to create new problems, or one might say, new and "better" problems. The attainment of any level of "success" of the human species always introduces more problems than it solves, but the problems are in some sense better because they are founded upon what has gone before. As viewed from the vantage point of the twentieth century, the problems of the working class in the nineteenth century were terrible problems of human deprivation, outrageous policies with respect to child labor, complete indifference to the health of thousands of people, and so on. And yet, in some sense, the problems of labor of the twentieth century are far more acute because people have learned more about the proper role of labor in society, the needs of health, recreation, family, etc. Because of all the "gains" we have made in creating a "better" society, we feel far more deeply today our inability to solve problems of the "quality of life" than did our forefathers of the nineteenth century.

Hence the Singerian inquirer pushes teleology to the ultimate, by a theory of increasing or developing purpose in human society; man becomes more and more deeply involved in seeking goals. To be sure, he may engage in relaxation, in playfulness and other forms of the non-serious, but he does so with the more fundamental purpose of re-creation. Comedy is the prelude or interruption of the heroic mission.

Can thought find the hidden ateleology of so strongly entrenched a teleology? The question is a very subtle one. One might argue that there can be a teleological defense of the Singerian ideals of science, plenty, cooperation, and the heroic mood. Yet one could let thought provide another twist by arguing that "defense" is per se a teleological activity.

The Backward Twist: The Teleology of Ateleology

Thought can continue on its twisting path by reversing the tables completely. Consider again the assertion that teleology is not to be extended to the pure feeling of value in the single act. There, for example, is the pure scientist, who takes his whole value to be in the satisfaction of his own intellectual curiosity when faced with a deep question he wishes to investigate in nature. It is absurd to argue that the satisfaction of his curiosity serves no other end. To be sure, that scientist at that time does not intend that his activity be a means to other ends. But he is not a separable component of society. Other members of society, e.g., the designer, have every reason to examine the more ultimate benefits of his acts. Society justifies its support of the ateleological attitude of the pure investigator, who wants only to satisfy his own curiosity, by judging the extent to which such activity seems beneficial to the rest of his community as well as to himself. The teleologist sees no conceptual embarrassment in saying that a man may pursue an activity for its own sake and not be willing to recognize any benefit that such activity may have beyond the happening or occasion itself. The feeling that is expressed by the philosophical ateleologist is therefore just that: a feeling to be explained by the very teleology of the living being.

Thought vs. Antiteleology

This is enough of the twisting path of thought. Ateleology fails to capture the spirit of antiteleology. It is too caught in its own teleological processes. It's all very much like the story of the planner who was having difficulty in persuading management to hire him because the managers were not sure they needed long-range planning. "All right," said the planner, "then hire me to help you plan whether to have planning!"

The real confrontation of antiteleology is with the thought processes of teleology. The basic confrontation occurs in the concept of uniqueness. The teleological *Weltanschauung* is like the deterministic and probabilistic in that its thought proceeds by using particulars and generals; it describes particular aspects of nature and can subsume all particulars into a more general class. Almost all planning has the form of particularizing people—that is, people are described as having incomes in such-and-such a range, home locations, religions, race, etc. Two people are the "same" for the planner if they have the same set of basic properties.

What shall the teleological designer say when the poet or the philosopher claims that the individual act of loving is unique and is not to be subsumed under a set of properties of biological reproduction, or psychological libido, or whatever? The experience that one individual has in his love for another individual is never duplicated, and, indeed, the concept of duplication simply destroys the whole feeling of love itself.

The theme of uniqueness has taken many forms. I especially like Kant's handling of the theme in his *Foundations of the Metaphysics of Morals* (1898), where he enunciates the moral law to treat every man as an end-withal, never as a means only. But every teleological plan I have ever seen is based on treating some, and perhaps all, people as means only because it is impossible to regard everyone as an "end-withal." Kant recognized this to be the case, and in modern language saw an eternal conflict between systems planning and morality; every plan must be partially immoral. The *Foundations* and more especially the *Critique of Practical Reason* (1788) are both forerunners of what I have called the Singerian inquiring system, for Kant believed in a gradual convergence of morality and systems planning or, as we would say here, between antiteleology and teleology. But a great deal more would have to be explored before we could understand how these conceptual enemies might ever become closer to one another.[1]

Thomas A. Cowan (1963) has applied the theme of uniqueness to the practice of law. The theory of law itself seems to have gone through some of the processes we have described of the inquiring system. There have been thoughtful attempts to understand the decision making of the judge in terms of generalizations which in part are the precedents of the legal profession. The particulars, or "facts," are taken to represent instances of these generalizations; the classical syllogism leads from the generalization plus the facts to the decision. There is no uniqueness here, and if one were to accept this model of the practice of law, one might be encouraged eventually to think that law itself could be subsumed under a teleological decision-making model. Perhaps a great deal of legal decision making could be handed over to some more or less automated process. But many lawyers and judges feel that in every case there is a unique element which co-determines the decision. Furthermore, the uniqueness of the decision is such that neither precedence nor the subsumption of facts under general laws can account for the particular decision that is made. What appear to the teleological analyst as "essentially" the same cases may be decided quite differently. To the analyst this may look like non-rational decision making, but for the lawyer it may appear rational because he believes in the underlying uniqueness of each decision.

Certainly many of the other professions, medicine, engineering, and teaching in particular, have often expressed somewhat the same philosophy, namely, that no amount of decision-making analysis can ever capture the unique properties of the great doctor, engineer, or teacher. No teaching machine can ever display the charisma that is associated with the inspired lecturer and his relationship to his students. No auto-

---

[1] I've struggled a bit with this problem in "Morality and Planning" (1969).

mated engineering design device can capture the particular unique ability of the great engineer to see in a flash the underlying essential problem to be solved. No automated diagnostician can capture that fine relationship existing between the patient and the doctor.

Of course the teleologist is not done, even if he takes the confrontation seriously. He points out that the scientist from the beginning has been told that his methods will forever fail to capture the true meaning of some set of phenomenal events. At a given time and for a given group of scientists there may be the strong feeling that the problems they face are essentially unsolvable. But time after time, science has achieved the kind of breakthrough in the pursuit of knowledge which permits it to understand what was once regarded as nature's eternal secret. To a generalizing mind, the assertion that the judge performs a unique event at the time he reaches his decision while on the bench is analogous to the assertion that the true distance between the centers of gravity of two planets at a moment of time is a locked secret of nature. To be sure, in some sense the true distance is a secret of nature. Such an event in astronomical history at a moment of time will never be completely understood by the scientist; there will always be some aspect of the event which his methods of measurement, no matter how fine, will fail to capture. There is no question that here is a unique event in the history of the world, this distance between the center of Mars and the center of Earth at the very moment of the beginning of this year. But this does not in any way imply the inappropriateness of the teleological model in the design of an inquiring system that is forever asking more and more about the same event. So if there is uniqueness in the moral individual and in the judge's decision, then let us pursue them with the same spirit with which we pursue all unique events of nature, and accept that they stand as limit points in our endless pursuit of more and more understanding.

The Inquiring System: Artificial or Natural?

The individualist, the poet, the man of today who is so frightened by science and its implications must ask himself again and again how the scientific enterprise can be stopped. The demands of the inquiring system go on and on; no aspect of nature, no problem that can be posed, ever blocks the ambitions of the designers of inquiring systems. Inquiry will enter into every domain of human life, into the deepest secrets of every human being—if need be, into the love between two individuals.

Is this a cancerous growth, this inquiring system that recognizes no end to the extent of its design, or is it simply a healthy development of the human race whereby, in an Aristotelian sense, we come more and more to be a *human* race? Does humanity find a true expression of itself through inquiring systems, i.e., through its understanding of the natural world in which it lives?

So calm and inspired a picture of science is very beautiful were it not for one thought so deep in its meaning as to make its expression almost hopelessly difficult. It is the thought inherent in a great deal of the discussion of this chapter: that the inquiring system, at least as its design has been conceived thus far, is not itself human, or rather, that what little humanity it has is simply an expression of one side of the human being, namely, the human being conceived as information processer and rationalizer.

We have in fact been discussing the design of the inquiring system, but what have we to say about the life of the inquiring system? If we take the life of the inquiring system to be inherent in nature, then no better topic for our final reflections on the design of inquiring systems could be found than their nature.

: 15 :

THE NATURE OF
INQUIRING SYSTEMS

Can Computers Think?

There will be many a reader who began this volume with the strong feeling that, despite the disclaimers of the author in the first chapter, the real intent of the book was to determine how one might use high-speed computer-processing machinery to perform the acts now commonly performed by the research scientist. Certainly almost everything that was said about the design of the Leibnizian inquiring system, for example, is at least conceptually designable into a program for a computer. Indeed, Chapter 4 provides one example. Similarly, rather simple Lockean inquiring systems have been represented in computer form. Perhaps one of the most fascinating problems in the development of so-called "artificial intelligence" is the simulation of a Kantian inquiring system, i.e., the system capable of applying a number of different models to a set of "inputs" in order to arrive at an adequate response to questions.

The ambitious designer might hope to continue beyond the Kantian systems, to develop the more complicated and more exoteric inquiring systems discussed in this book. Of course, at the present time many so-called automated systems, e.g., management information systems, are essentially stupid. They are completely incapable of changing a useless pattern, once the pattern has been set into them. They cannot be said to think in depth; at best, all that they are capable of is an elaborate manipulation of data storage banks.

Nevertheless, we are only beginning to design automated systems to assist us in our inquiries. It is probably a futile argument to speculate on whether a computer system is "intelligent." They are already capable of making high scores on certain kinds of intelligence tests. Thus Persson (1966) developed a Kantian-like program for sequence extrapolation (e.g., given 2, 3, 4, 9, what is the next number?). Persson's program seems to be fantastically intelligent in this domain, for it can take a scrambled sequence, unscramble it, and extrapolate the next number. What is most significant about its success, however, is that once you understand how it works, its thinking appears trivial. The program contains several alternative schemes for generating sequences (polynomials, a simple Markov process, a binary symbol scheme, etc.); given a sequence, it tests to see whether one of its schemes fits, and if it does, it uses it to generate the next symbol. Thus its ability to unscramble is "obvious," as Watson would say to Holmes once he understood Sherlock's method.

The fact that the computer program becomes trivial or stupid once we see how it works has some very important implications for testing the intelligence of humans. Suppose a young boy has a fantastically high score in an intelligence test; if we knew his program, would he then appear stupid? Surely if he simply had a few tricks like Persson's program and an idiot-savant capability of memorizing and computing, we would not judge him to be intelligent. Why? Because intelligence is a natural quality of humans, and mere computation does not capture the essence of this nature. Thus instead of asking whether computers are conscious or intelligent or capable of thinking, it is far more fruitful to ask in what way people are capable of being conscious, intelligent, and thoughtful.

Do Scientists Think?

Our walk in the forest will now become meandering again as we try to get a glimpse of the meaning of "natural." Suppose we start with a more or less playful mood, and suggest that "natural" means "normal for the species." The deer in the Kyoto park are unnatural because they don't bound away when you approach them; Saki's Tobermoroy was unnatural because cats don't normally talk out loud. This suggestion is playful because we can use it to make fun of the scientists, especially the more brilliant of the crowd: they are obviously unnatural. After all, very few members of the human race ever attain the status of being even mediocre scientists. This does not say, of course, that the scientist is completely unnatural; he may be a natural lover, for example. But it does imply that it is unnatural for the human being to be a scientist; it does not come out of the nature of most human beings to be inquirers in the sense that the scientific community recognizes inquirers. This is a bit of fun with a serious twist: the common fear of the scientist by the layman is all too natural.

The Natural Functions of Mind

But enough of that. To become more serious, we take another pathway, which begins by recognizing that the inquiring system is a mind. We then ask ourselves what is natural about minds. To explore this question in any depth would of course take us down many pathways, but instead of following all of these, suppose we retrace one of them, namely, Jung's theory of the psyche. In his *Psychological Types* (1959) Jung classifies the functions of "mind" into thinking, feeling, sensation, and intuition. Jung's idea was that in the "normal" development of mind certain functions may become more developed than others, in which case the less developed functions act as part of the "unconscious" mind of the individual. Superimposed on this four-way classification of the mental functions was Jung's more famous introvert/extrovert classification of temperaments, the extrovert presumably being the individual who is more outgoing in his energies, the introvert the one who finds his life in the inner world.

Jung further suggested a dynamic model of the functions. For example, if thinking is a highly developed function of the individual, then feeling will be the underdeveloped function. Similarly, if feeling is the developed function, then thinking will tend to be underdeveloped. A similar relationship exists between intuition and sensation. The schema is filled out by combining the two temperaments with the functions, so that extroverted thinking will have introverted feeling in its dark side.

Examples of the Jungian functions can easily be found in the designs of inquiring systems in this book. For example, the Leibnizian inquiring system seems to be a "thinking type" in its orientation; its whole construction begins with a particular kind of thought process. What Spinoza tried to show is that thinking cannot occur without intuition. Sensation, the ability to discriminate finely, is clearly required in Leibnizian as well as the Lockean inquiring systems. In Lockean and Kantian systems, intuition plays the role of creating the right kind of generalization or the right kind of model to apply to the data. In the Hegelian inquiring system the interplay between the three functions of sensation, intuition, and thinking becomes even more involved. In the dialectical process intuition is the force that breaks through the apparently insuperable conflict of ideas to create a new position which can observe the conflict and pass beyond it. It is "natural" wherever intuition plays such a role for thinking to try to take over and understand the operations by which intuition was so successful. This is one characteristic of the thinking function—to transform the success of the other functions into thinking. Such transformations have been discussed repeatedly in this book in the various suggestions that have been made as to how the inquiring system might make its generalizations, or seek for appropriate models to interpret its data, and so on. Thus intuition is always important in the development of the mind of the inquiring system, but the challenge to the thinking designer is to rationalize the operations of intuition, so that the creativity of one man becomes simply the methodology of another; the great idea of one generation becomes the mundane operating basis of the following.

Do Inquiring Systems Have Feelings?

What, then, is a "natural" mind? The implication of Jung's theory is that "natural" means "developing"; a mind is natural to the extent that it comes in touch with its unconscious as well as its conscious. If its feeling function is underdeveloped, then it is natural as the mind ages that it bring its feelings into consciousness and work through the feeling function. It isn't important whether we use Jung or Freud or some other contributor to the theory of mind; what is important is the idea that a natural mind is a mind that seeks a certain kind of completeness. In his *Two Essays in Analytic Psychology* (1953) Jung states the theme forcefully: ". . . the man who is pauper or parasite will never solve the social question." "Pauper" is not an economic description, but rather refers to the man possessed of "blinding illusions" because he has failed to seek his own individuation.

Well, then, are the inquiring systems of the first part of this book "paupers" or "princes"? Anyone can see from our recent discussion of the systems that feeling was neglected; no new story emerges for this second part. To the thinking type, Jung's own description of this function in *Psychological Types* is frustratingly vague. Sometimes it seems merely to mean a sense of appropriateness: it is inappropriate to defecate at the dinner table.

We have already seen how deeply the inquiring systems of Part I go into the topic of evaluation in the more serious sense of what is ultimately valuable for the human being. For example, in the rationalist inquiring systems of Descartes and Spinoza, great attention is paid to the problems of evaluation; indeed, in their designs the ultimate basis of evaluation must be established before the system can be thought to have any meaning whatsoever. The ultimate basis lies in God, and the system of evaluation follows from the properties of God. Nevertheless, this concept of evaluation is apparently not what Jung had in mind, since it in no sense represents a contrast with thinking. On the contrary, to rationalists like Descartes and Leibniz, the ability to prove the existence of God depends essentially on the thinking function or, in the case of Spinoza, on intuition.

Lockean inquiring systems design the evaluation function by a strong reliance on sensation. For example, accompanying certain sensations there is an internal sense of pleasure or pain. The "evidence" of good and bad lies in the particular reaction of the organism in its sensory life. Out of such a preliminary design emerges the modern idea of evaluation in terms of maximization of utility.

An interesting recent development of the extension of sensation and thinking to evaluation is the application of research to the evaluation of the research-and-development process (so-called R^2, for research-on-research). These studies attempt to determine how research is conducted, and how it is organized and administered, especially in western culture. One aspect of this effort develops economic models of the research-and-development process, so that the policy makers can make appropriate judgments about the allocation of funds to various kinds of research projects. They can also determine at what point projects should be dropped or extended beyond the original plan. These "cost-effectiveness" studies therefore try to apply evaluation from a social point of view to the inquiring system itself. As we have already pointed out, there are enormous difficulties in trying to apply classical economic or operations-research models to the research-and-development process. The enthusiasts, however, claim that eventually the evaluation process will be successfully implemented. They might go so far as to say that they will successfully design the feeling function of the inquiring system.

In sum, the thinking designer has worked very hard to extend the functions of the inquiring system to "take care of" the missing function in the Jungian scheme, i.e., the evaluative function. It is certainly fascinating to see how feeling becomes the handmaiden, so to speak, of one of the clearly recognizably useful functions of the inquiring mind.

The Feeling of Feeling

Nothing in our walk is conclusive. But we can appreciate why the thinking designer would try to protect himself from the suspicion of incompleteness: once the "other" image becomes a possible reality, the whole of his design may collapse.

To understand what this means, imagine an inquiring system which says *not,* "It is true because it is so," but, "It is true because it is beautiful." A story about two people which reveals the beautiful depths of their love for each other is true; the story that recounts their fierce fights is merely so-so.

Could the beautiful ever be the basis for the true? Should the designer of inquiring systems design them to be beautiful? If so, does he dare to be so outrageous as to figure out, analyze, think about the components of the beautiful? Shame on him! He'll be doomed to the limbo of those thoughtful philosophers who invented that most dreadful field of inquiry, aesthetics. Perhaps, too, shame on him for thinking he could design a religious inquiring system, or one that probes the depths of the unique individual.

But shame is a feeling word, not a scientific concept. Did Galileo's prescription to make all things measurable include the prescription to measure shame: "You are twice as shameful as I"?

The Unconscious Side of Science

If it is true that science consciously makes thinking its dominant function, with intuition and sensation as subsidiaries, then, assuming that Jung's theory works for inquiring systems, its unconscious side is feeling. What is more "natural" than science's attempt to suppress its unconscious side by designing a surrogate for it, a very common psychological phenomenon.

In an attempt to describe the unconscious of the inquiring system we might try the following idea. Suppose we say that the unconscious represents those aspects of the inquiring system which the designer is incapable of designing. In this sense, the stream of inputs to a Leibnizian, Lockean, or Kantian inquiring system is in the "unconscious mind" of the system because the designer of a system does not design the input stream. The idea that the inquiring system is an input/output system is itself part of the unconscious mind of the inquiring system, since in many designs this assumption is a given, i.e., is not subject to the control of the designer.

In Hegelian inquiring systems, with their capability of looking at the inquiring system as a whole, the unconscious mind of the earlier designs becomes conscious. Now the entire system, including the observing system, becomes aware of what "input" and "output" mean; further, the designer has the capability of deciding whether an input/output schema is the most appropriate one to apply to the design of an inquiring system. We might say that as we moved from the Kantian to the Hegelian design, certain aspects of the sensation function were brought up to consciousness.

Nevertheless, Hegelian inquiring systems are only partially capable of observing themselves. As the discussion of implementing and religious inquiring systems showed, the colossal problem of a mind's being able to understand itself requires that the mind go far beyond its own so-called boundaries to incorporate more of the world, both broadly and deeply. Indeed, the implementing inquiring system in some fundamental sense seems always incapable of really understanding itself. Wherever these incapabilities occur, one could regard them as aspects of the unconscious mind of the inquiring system.

If our walk were of the terminal type, we might conclude by saying that if science, and particularly the designers of inquiring systems, took the lessons of religion and psychology seriously, then it would be apparent that self-analysis is called for. The central task of the designer is to come in touch with and work through the unconscious of the inquiring system.

"Natural" Is Differentiative

But our walk is more like the behavior of the blue-bellied lizard. Some teleological scientists became curious about the activities of these creatures; surely their jutting about had some purpose. Perhaps those who jutted about more survived more. But they sadly or gladly concluded that the blue-bellied lizard is just plain fooling around 50 percent of the time.

So let's try another meander about the "natural." This one begins by considering some other functions of the human being, namely, sex and reproduction. Practically every human being is capable of expressing these functions. As he becomes an adult he begins to express them in his own style. It may be true that history records experiences of great lovers who had some kind of marvelous or astonishing ability in their love making, but none of this in the least detracts from the particular expression that an individual human being may have of his own sexual experience and its relationship to those he loves. Nor would it be "natural" to feel inclined to go among the members of the human race giving them ranks as first-rate, second-rate, third-rate lovers.

When we turn to another function, nourishment, the situation changes somewhat. Here again, every person is capable of expressing this function, often in his own particular style. The style may be based partly on national fads or cultural values, but usually has components of his own individuality in it. Of course it is true that various cultures develop criteria of excellence in cookery; they recognize those who can make fine distinctions between various kinds of tastes, so that a culture might be perfectly willing to say that a certain cook is "first-rate" compared to the rest of the cooks of his society. But in a way this ranking does very little so far as the individual and his style of nourishment is concerned. If he dearly loves English boiled beef, boiled potatoes, and (very) boiled Brussels sprouts, then it matters very little to him that some gourmet exists in Paris who is able to distinguish between two excellent years of vintage in wines. His own tastes and expressions are his and represent part of his own individuality. Another may recognize him as a rather dull individual from the point of view of his sensory life, but this recognition is quite irrelevant in the individual's natural expression of himself. To be a gourmet of cookery doesn't imply that you eat better, whatever the French may say.

But when we turn to what are called the higher functions of our culture, e.g., the arts, the matter becomes more difficult. Western culture has been quite willing to recognize fine distinctions between the quality of output of various artists and writers. Nevertheless, art is very common and natural among humans. Most people express themselves in one way or another in art forms: painting, singing, piano, dancing, play acting, whatever. Anyone who has enjoyed the experience of sharing his art experience with another comes to realize that he is little concerned about whether his art object will attain the dizzy heights of the finest productions of the age. No matter that the painting is buried in the attic, that the song goes flying forever into the air, that the dance of an evening is over and done with, or that a play-acting piece is enjoyed by a few victims of an evening. In all these cases the individual has come to recognize an expression of himself, and the functioning of this expression is clearly what matters.

None of this says that teaching, training, and other sorts of help are inappropriate for the individual. They may be highly appropriate. He may find that by going to art classes he develops a much more satisfying way of expressing his own nature. In other words, the "designer" may have a function to play in all the ways in which people express themselves artistically. As leisure time becomes greater in western society, we may expect that many individuals will find careers in helping individuals by developing their functions, sexual, gastronomic, or artistic.

From the point of view of an archivist, or of one who tries to restrict publication to only the finest, this development may be disturbing. He envisages, pouring out of the increased hours of leisure time, a flood of millions of pictures, songs, novels, choreographic patterns, and so on, which must be assimilated in some way by the society, either destroyed or placed in various kinds of archives. The person searching among this mass of output for the "best" may find his life far more difficult than it is at the present time because of the necessity of examining so much.

But from the designer's point of view the archivist's concern may be a matter of far less importance than providing opportunities for people to discover forms of expression appropriate to their nature. After all, keeping archives is just another expression of individual style.

What shall we say about inquiry? Why, simply that it, too, is one of the most natural expressions of the human race. Everybody has within him the spark of curiosity, the need to differentiate between what is more likely to be true and what is more likely to be false. Every individual feels the wonder of the stars above him, the magnificence of the expression of life in the plants and animals that wander about his world, the nature of the depths beneath the sea, the patterns of rocks and earth upon the land, the nature of his own mind as well as that of his friends, and, of course, the nature of the political debates that abound in the world today.

Yet very few people would recognize themselves as amateur scientists, nor would they recognize many moments of the day in which they could say that they had spent a part of their leisure time in inquiry. Few are inclined to keep laboratories in their basement or spend some of their weekend hours wandering the country observing different kinds of flora and fauna. There are a few who choose amateur science as a hobby; they probably feel that the findings of the amateur scientist need not be the highest level or even publishable in one of the recognized dignified journals. The joy of inquiry is enough.

But most people would find the idea of spending leisure time in scientific inquiry to be completely beyond their capabilities, particularly because of the necessity of devoting so many years to intense training to prepare themselves. Instead, in our culture the only expression of himself a man can find in scientific inquiry is to read one of the many hundreds of popular books that explain various scientific results. But then he becomes simply the passive recipient of the activity of inquiry, much as the appreciator of art who goes to a museum is often a passive recipient of the creativity of another artist. The ordinary man never comes to feel the act of discovery as part of his own natural life.

But it would be foolish to say that most people are indifferent to inquiry. They are curious about all the important things of their life— their family, friends, neighborhood, nation, and now, more and more, the "ecology" of their outer life and the meaning of their inner life. But their style of inquiry is not that of a scientific discipline. It is as much their own style as it is a common method. Nor should anyone pose as an authority on how a person should inquire, although many people can help others develop their style of inquiry. To paraphrase an old saying: "An inquirer is not a special kind of person; rather, every person is a special kind of inquirer."

So "natural" means "in accordance with one's own nature." And "education" means "leading oneself into one's own expression of inquiry." If we were on a teleological walk, we could say a great deal about the implications of this meaning of "natural" relative to modern education. There are many compelling reasons for saying that the design of modern education from grade school to Ph.D. is seriously faulty. So-called mass education is a device for suppressing the natural in the student, not nourishing it.

The Exoteric as Science's Unconscious

The last pathway meanders back to the one before, to science and its unconscious. Science has learned a great deal about the social life of man, his health, environment, technology, and so on. This learning, which is a style of inquiry, makes the scientist an "expert." Once this title is conferred, it is but a step to the assumption that the expert knows best, that his style of inquiry in matters of social policy is better than other styles. So he wonders why his recommendations are not implemented.

We might then say that the unconscious of science is what "everyman knows" in his own individual way. There are no experts in inquiry. Exoteric inquiry may be based on esoteric disciplines, but it may also be based on all sorts of other experiences, of a life of love or pain or poverty or adventure.

Thus the natural act of acquiring understanding about the world in which we live is highly differentiated. But our society is not well designed to permit the differentiation to unfold. One of the more tragic and also ridiculous aspects of our society is that the area in which men most urgently need to develop their individuated forms of inquiry is an area where society tends to mold them: the political and economic world. The *Weltanschauungen* of politicians, newspapers, and magazines are all similar, even when they appear to be at odds with one another.

We hear a great deal about the "well-informed" public. The concept of a well-informed public seems to imply that each mature adult citizen will have acquired the status of at least an adequate inquiring system with respect to the political problems of his society. But this is absurd. No one can acquire such a status. Rather, a well-informed citizen should be a person who has been able to mature his own individuated mode of inquiry. Everyman knows that no one knows how social policy should be formulated.

In Defense of Expertise

In our pathway is the scientist again. He doesn't take kindly to the presumption that we know his unconscious. Furthermore, he reminds us, we have failed to make an important distinction (he's always reminding us about this!). Everyman may be somewhat of an expert on what he wants in life; the scientist does not mean that he can tell people what they should want. But the scientist is an expert in some very important cases on how society may provide the means for attaining what people want. For example, consider the question of building a dam on a river. If the dam prevents flooding and supplies a water resource, and the people want flood prevention and water, then they need experts: experts on dam construction, financial resources, design of new communities, minimization of damage resulting from the construction of the dam, and so on. The ordinary citizen cannot be expected to inquire into these matters; he has neither the background nor the time.

The same distinction applies to the other critical policy issues of our society: nutrition, population, war, communication. Too often, says our scientist, the citizen is asked to vote on a deeply complicated policy which depends on many technical issues he knows nothing about (as in bond issues, say). No wonder that the typical inquiring system of the citizen of the United States tries to find some ready categories into which issues can be placed. For example, an uprising in a given country will be taken as a sign of "communist infiltration" because communist infiltration is a matter easily to be explained in terms of the aggressiveness of one nation whose ambition is to control the world; the answer to communist infiltration seems obviously simple, namely, to try to stop it. Therefore, the frustrated citizen tries to translate a difficult political-economic problem into an emergency-type problem, where there is only one obvious pathway to be taken; the rest of the information and analysis becomes irrelevant. You talk of "natural inquiry," says the scientist; why, this way of translating all of the critical social problems of our day into problems of emergency is most unnatural. It flies in the face of reason and also of feeling, because it characterizes certain peoples of the world as "bad guys" while other people, ourselves, are "good guys."

Distinctions Are the Weapons of the Elite

The response of the layman is a cry of anguish. "You destroyed my valley with your dam. Did your expertise tell you what the valley meant to me, to my relationship to my family and my God? You counted desires, those who desired water and those who desired the land, and your ridiculous count said a thousand-to-one. Is this what expertise means? Suppose I'd had the million you had to build the dam. Suppose I'd used it to go into the homes of the water users, to tell them what my valley meant to me and my friends. Would your absurd count have then been a thousand-to-one? Do you really think you can count human feelings?"

And so on. Nothing will ever stop the scientific Lockean community from making distinctions, and nothing will stop others from ignoring them. The designer, however, must ask himself whether a distinction implies a separation of components of the system, and if so whether the separability makes sense in the design. We have every reason to suspect that the distinction made above leads to an unsatisfactory separability of the components of the social system.

In the chapter on dialectical inquiring systems we were searching for a way in which society might aid people to develop their own style of inquiry, namely, in terms of a well-designed debate. In this case a person's own internal feelings might enable him to judge whether the story which is told to defend one policy is a better or worse one than the story that is told to defend another policy. In time and with practice he might himself begin to develop in his own style the particular picture of the world as a whole that is most appropriate for him. At such a time we might be getting nearer to that point in history where inquiry becomes the "natural" expression of the human being again. But of course the dialectical debate was itself a product of a science of some sort, and the sanguine mood of its inventors is their expression of their own faith.

Some Lessons

Obviously, we haven't been meandering at all. What seemed like an aimless walk was the direct pathway to some teleological lessons about the design of inquiring systems.

For example, we have learned that attempts to make the intelligent population "appreciate" science are ill advised if they merely consist of explaining what scientists do and how their results are used or not used in the culture. Such educational courses and explanations fail to touch the real nature of inquiry from the point of view of the listener. If we are to learn from the last chapter, in the creation of natural inquiring systems the individual himself must have that particular feeling of action-and-change which was demanded in the antiteleological schema. In some sense, then, inquiry must happen within him and not be a happening that he merely observes.

Next, we have learned that some steps in the direction of developing natural inquiring systems could be taken if men and women in their various activities could come to see in what ways these activities could be construed as inquiry. For example, the manager of a corporation normally thinks of himself as the administrator or policy maker. He is not accustomed to thinking that what he does is to inquire, that his various actions might be construed not as maximizing the profits, but as discovering something about how a particular organization behaves in a certain type of environment when certain things happen to it. He might be aghast at the notion that each decision on his part could be construed as an "experiment" because to experiment with the funds of a corporation might seem irresponsible. But if he had a broader insight into the nature of inquiry, he might come to see that this description is indeed appropriate; each action on his part is an attempt not only to improve the financial status of the corporation, but also to increase the understanding of the way in which the corporation behaves and ought to behave. If the designer were to intervene at this point, he might assist the manager in becoming more effective by pointing out ways in which the natural expression of his inquiring function can be developed.

In the same way, the busy politician intent on the problems of deciding how to vote on various pieces of legislation may fail to realize the manner in which his own activities become expressions of his inquiring function. Of course, nowadays the politician often makes use of so-called "hearing committees," which have some of the flavor of inquiry about them. No doubt many politicians believe that hearings are the basis of their inquiry. For them to realize that the total function of politics is to inquire into the nature of the state and its policies might very well improve the design of the hearings. And at the level of Everyman there ought to be a way in which he begins to understand that inquiry is a central part of his psyche, that it is natural for him to give expression to this function, and that in many cases it is natural for him to try to learn about the function in order to make his own individual expression of it more complete.

Returning now to Jung and the unconscious, it may be safe to say that at the present time inquiry has become part of the unconscious life of most people: they are unaware of the ways in which they function as inquiring systems. Nor is there a strong inclination for them to give expression to this function so that its nature appears at the conscious level. As a consequence we are suffering now the most dangerous symptoms of an inability to bring to the conscious level an important human function. We assign to the experts and the politicians the roles of designing and creating the environment in which we live because we can see no way in which we can play any role whatsoever in these activities. Appalled as we may be at the events that are occurring in the world about us as the output of blind technology and politics, we each in our own frustrated way feel that we can do nothing about it. We cannot recognize ourselves as inquirers, i.e., as capable of understanding what is going on or of implementing any understanding that may occur.

The very word "science" as we use it in present-day life is symptomatic of the sad state of the natural inquiring system. We believe that people can only become scientists by means of certain kinds of training through the Master's and Ph.D. levels, where they are required to take specific courses for which their professors have no justification. No one can claim that these courses create the scientific mind. We recognize as "scientists" people with the most narrow perspective of the world in which they live, who confess that their chief interests lie in trying to understand only one segment of the world, and whose particular mode of inquiry is a rigidity brought about by a strong mutual agreement among the leaders of the scientific community.

Those who love science, in both the esoteric and the exoteric sense, must also be its fondest critics. The true patriot today is the man who recognizes the deepest weaknesses as well as strengths of his beloved country because this kind of recognition is the essence of love.

The real problems of "science" today therefore do not lie creation of new scientific discoveries that constitute so-called breakthroughs in various disciplines of science. These discoveries no doubt will continue to occur and will fascinate and please the human mind that has a curiosity concerning them, but the design problem of science in these areas is of minor importance. What is most important of all is to create natural inquiring systems, i.e., the expression of the activity of inquiry rather than its mere appreciation.

These at least are some lessons for our designer. But what of him and his unconscious? If we speak of "designing" natural inquiry, haven't we merely relegated the evils of the unnatural to the designer? Can anyone but God design nature?

: 16 :

THE DESIGN AND NATURE
OF INQUIRING SYSTEMS

The Guarantor and the Natural

We began the speculation of this book on the design of inquiring systems with the formulation of the critical problem of all inquiry, namely, the nature of the guarantor of the validity of the results of man's attempt to gain knowledge. We have ended with an equally critical problem, namely, how inquiry can become natural. This is not a book on "scientific method" in that it is not chiefly concerned with the methods whereby science has attempted to discover and verify. It is rather a book on the way in which philosophy plays a role in the interpretation of man's attempt to understand his world.

The critical problem with which we began, so simply stated by Descartes, was how an inquiring system could possibly have any meaning in terms of establishing valid results unless it had some firm concept of the whole system and the way it was governed. Did anything that followed this earliest question of the essay assist in answering the Cartesian question? In some ways the answer must seem to be no. We have shown how the complexity of the inquiring system must be increased and the designer must be faced with more and more perplexing problems in his attempts to cope with the structure of the system and its relationship to other activities in its social world. And yet we found no clear guide to a principle of guarantee, except in the end to say that it seems to lie in the nature of man. The faith that seems to become so necessary once we realize that inquiring systems are implementing, i.e., have a relationship to their world, arises out of the particular nature of the inquirer himself. But in the end we have come to see that we have no clear idea of the relationship between the inquiring function and the total mental nature of man.

In another time or in another culture, I might have entitled this last chapter "God and Nature," thereby expressing the classical notion that the nature of man lies in his God—that the problem of the designer of the inquiring system is to recognize God for what He is and to worship Him for the particular characteristics He has which are so relevant to man's own being. One might then have come to feel that the questions raised in the last chapter concerning the nature of mind and the way inquiry expresses this nature are very much like the question about the nature of God and the role He plays in inquiry.

Inquiry Revisited

Perhaps the two questions, the one concerning the guarantor of the inquiring system, and the second its nature, can be closely related in terms of the central problem that has concerned us throughout the book, which is the meaning of inquiry itself. As each design has been discussed, the meaning of inquiry has changed. If a designer is satisfied with the input/output conception of the inquiring system, then he defines inquiry as some kind of processing of an input stream which produces an output with certain desirable characteristics. Once the designer faces the problem of the implementing inquiring system, on the other hand, he has to think of inquiry as playing a role in the way in which man makes his decisions and attempts to control his environment. And to the antiteleologist, inquiry becomes a "feeling of knowing," a kind of activity which the human being can express, and in its expression find the only real value there is to that mode of living.

In the teleologist's model, inquiry is precisely defined by a classification of the different kinds of goals that men seek. Inquiry becomes the creation of an ability of the human being to solve his problems, to discern better pathways to goals no matter how the environment may change. In nineteenth-century terminology this was called the "control of nature," not a happy phrase, to be sure. To control nature is to be unnatural.

In the disciplines of science each discipline expresses the meaning of inquiry by means of its techniques and methods. Outside the disciplines there can be no universal guides to the meaning of inquiry.

Inquiry is the creation of knowledge or understanding; it is a reaching out of a human being beyond himself to a perception of what he may be or could be, or what the world could be or ought to be.

If, in this broader and far vaguer notion of inquiry that has appeared in the later pages of this book, one asks now about the guarantor of the inquiring system, the same kind of vagueness will naturally result. What is the guarantor of man's nature?

Design and the Natural

At the outset in Chapter 1 it no doubt appeared that design and nature were antithetical. We asked what aspect of an inquiring system cannot be designed, and the flavor of the question seemed to be: What aspect of inquiry cannot be made artificial, e.g., programmed on a computer? But now design is allied to conscious: What aspect of an inquiring system cannot be made conscious? This is the "natural" question about design.

Hence design, too, is a natural function of all of us. It is one that is expressed in the many activities that we engage in If the teleologist has his way, design means the conscious attempt to create a better world. If the antiteleologist has his way, then design is simply the conscious part of action itself.

And yet this statement has a relativist tone to it, as though the final answers to the puzzling questions we have been considering were to be found to rest with each individual himself according to his own lights. The rationalist's search for a guarantor was non-relativistic. Descartes was not in search of a principle that might satisfy this individual or that, nor would he have been satisfied to have his question answered by each individual in accordance with his own nature. No, the Cartesian guarantor must be general, applicable to all inquiry. Yet in the last chapter, as we began to explore the idea that out of the nature of each man there might be an expression of his inquiring function, we seem to have introduced a tone of relativity, namely, that each man should conduct inquiry according to his own nature. This, as we pointed out, flies directly in the face of the traditional disciplines with their high degree of control.

We have come, as might be expected, to the expression of the philosophical problem of this century, namely, the relationship between relativism and non-relativism. The optimism of the nineteenth century about the gradual progress of science was a clear expression of the non-relativistic nature of science—science was on its way to truth; the confidence of the scientist of that century captured the spirit of the Cartesian demand for the ultimate guarantor. The despair and doubting of men of this century concerning the value of the knowledge created by science, and the need of each man to express himself as he is, generate the relativist position in many, many forms.

And yet in the Jungian analysis there *is* such a thing as the nature of man; one can infer that the nature of man in the Jungian sense is non-relativistic. Man expresses himself in his own individual way, to be sure, and his concept of the guarantor comes out of his own individuality; yet his own individuality is a reality, and not relative to this or that inquirer's view of the world.

Conclusion

What conclusion can be drawn from this discussion? Is there a real, objective guarantor, or is his existence a matter of personal belief? Can the "creative" be designed; can "nature" be designed? If so, will the design really produce improvement, or will the improvement be relativistic or illusory?

"Conclude" comes from the Latin *concludere*, meaning "to shut up together"; in one sense it means a Lockean community which agrees to shut up. This inquiring system dissents; in the words of one of my philosophical mentors, Henry Bradford Smith, the only conclusion of philosophical discussion is a question. So: What kind of a world must it be in which inquiry becomes possible?

REFERENCES

ACKOFF, R. L. 1957. *Design of Social Research*. Chicago: University of Chicago Press.

——. (ed) 1961. *Progress in Operations Research*, Vol. 1. New York: John Wiley & Sons, Inc.

——. 1962. *Scientific Method: Optimizing Applied Research Decisions*. New York: John Wiley & Sons, Inc.

BENNIS, W. G., and SLATER, P. E. 1968. *The Temporary Society*. New York: Harper & Row.

BOULDING, K. E. 1956. *The Image*. Ann Arbor: University of Michigan Press.

CAMPBELL, J. 1956. *Hero with a Thousand Faces*. New York: Meridian Books.

CHOMSKY, N. 1968. "Language and the Mind," *Psychology Today* (February).

CHURCHMAN, C. W. 1961. *Prediction and Optimal Decision*. Englewood Cliffs, N.J.: Prentice-Hall, Inc.

——. 1968. *Challenge to Reason*. New York: McGraw-Hill Book Co.

——. 1969. "Morality and Planning," Internal Working Paper No. 98. Social Sciences Project, Space Sciences Laboratory, University of California, Berkeley, Calif., November.

——, and BUCHANAN, B. 1969. "On the Design of Inductive Systems: Some Philosophical Problems," *British Journal of Philosophical Sciences*, 20, pp. 311–323.

COWAN, T. A. 1948. "The Relation of Law to Experimental Social Science," *University of Pennsylvania Law Review*, pp. 484–502.

——. 1963. "Decision Theory in Law, Science and Technology," *Rutgers Law Review*, 17, p. 499; and *Science* (June 7).

——, *et al.* 1965. "The Legal Structure of a Confined Microsociety—A Report on the Case of the Penthouse II and III," Internal Working Paper No. 34. Social Sciences Project, Space Sciences Laboratory, University of California, Berkeley, Calif., August.

DANTZIG, G. B., and WOLFE, P. 1960. "Decomposition Principle for Linear Programs," *Operations Research Journal*, 8 (January), pp. 101–111.

ELLUL, J. 1964. *The Technological Society*. New York: Alfred Knopf.

FEIGENBAUM, E. 1963. "The Simulation of Verbal Learning Behavior," in E. Feigenbaum and J. Feldman, *Computers and Thought*. New York: McGraw-Hill Book Co.

FISHBURN, P. C. 1964. *Decision and Value Theory*. New York: John Wiley & Sons, Inc.

FLOOD, M., and LEON, A. 1964. "A Universal Adaptive Code for Optimization (GROPE)," Internal Working Paper No. 19. Social Sciences Project, Space Sciences Laboratory, University of California, Berkeley, Calif., August.

GELERNTER, H., HANSEN, J. R., and LOVELAND, D. W. 1963. "Empirical Explorations of the Geometry-theorem Proving Machine," in E. Feigenbaum and J. Feldman, *Computers and Thought*. New York: McGraw-Hill Book Co.

GERSHENSON, D. E., and GREENBERG, D. A. 1964. *Anaxagoras and the Birth of Physics*. New York: Blaisdell Publishing Co.

GOODMAN, N. 1965. *Fact, Fiction and Forecast*. Indianapolis, Ind.: Bobbs-Merrill.

HANSON, N. R. 1961. *Patterns of Discovery*. Cambridge, Mass.: Harvard University Press.

HELMER, O. 1966. *Social Technology*. New York: Basic Books.

HILLMAN, J. 1968. *"Senex* and *Puer:* An Aspect of the Historical and Psychological Present." Offprint from *Eranos-Jahrbuch*, XXXVI (1967). Zurich: Rhein-Verlag.

HITCH, C. J. 1965. *Decision Making for Defense*. Berkeley, Calif.: University of California Press.

HOOS, I. 1969. *Systems Analysis in Social Policy*, Research Monograph No. 19. London: The Institute of Economic Affairs.

———. 1970. "Systems Analysis as a Technique for Solving Social Problems—A Realistic Overview," *Socio-economic Planning Sciences*, 4, No. 1 (March), pp. 27–32.

HUYSMANS, J. H. B. M. 1970. *The Implementation of Operations Research*. ORSA Series No. 19. New York: John Wiley & Sons, Inc.

JAMES, WILLIAM. 1916. *Varieties of Religious Experience*, New York: Longmans, Green & Co. [c. 1902].

JUNG, C. G. 1959. *Psychological Types*. New York: Pantheon Books.

———. 1953. *Two Essays in Analytic Psychology*. Bollingen Series XX, p. 223. New York: Pantheon Books.

KANT, I. 1788. *Critique of Practical Reason*.

———. 1898. *Fundamental Principles of the Metaphysics of Morals*. Translated by T. K. Abbott. London: Longmans, Green & Co., Inc.

LEDERBERG, J. 1964, 1965. DENDRAL—64, Part I and Part II. NASA Reports CR–57029, CR–68898.

LEIBNIZ, G. W. *Discours de Metaphysique*.

———. 1914. "Réponse de M. Leibniz aux reflexions continue dans le second édition du Dictionnaire critique de M. Bayle." Reprinted as "Zweite Schrift gegen Bayle" in G. W. Leibniz, *Ausgewählte Philosophische Schriften*, H. Schmalenbach, ed. Leipzig: Felix Wemir.

LINDBLOM, C. E., and BRAYBROOKE, D. 1963. *A Strategy of Decision Policy Evaluation as a Social Process*. Glencoe, Ill.: The Free Press.

LOCKE, J. 1690. *Essays Concerning Human Understanding*, George W. Ewing, ed. Chicago: Henry Regnery Company.

MARSCHAK, J., and RADNER, R. 1971. *Economic Theory of Teams*. Monograph. New Haven: Yale University Press, Cowles Foundation.

MASON, R. O. 1968. "Dialectics in Decision-Making: A Study in the Use of Counterplanning, and Structured Debate in Management Information Systems," Internal Working Paper No. 87. Social Sciences Project, Space Sciences Laboratory, University of California, Berkeley, Calif., June.

MICHAEL, D. N. 1968. *The Unprepared Society*. New York: Basic Books.

MILL, J. S. 1862. *A System of Logic*. London: Parker, Son & Brown.

MITROFF, IAN. 1971. "Mythology of Methodology." To appear in *Theory and Decision*.

NEWELL, A., SHAW, J. C., and SIMON, H. 1963. "Chess Playing Programs and the Problem of Complexity," in E. Feigenbaum and J. Feldman, *Computers and Thought*. New York: McGraw-Hill Book Co.

————. 1963. "Empirical Explorations with the Logic Theory Machine: A Case Study in Heuristics," in E. Feigenbaum and J. Feldman, *Computers and Thought*. New York: McGraw-Hill Book Co.

NICOD, J. 1917–1920. "A Reduction in the Number of Primitive Propositions of Logic," *Proceedings of the Cambridge Philosophical Society*, Vol. 19.

PERSSON, S. 1966. "Some Sequence Extrapolating Programs: A Study of Representation and Modeling in Inquiring Systems," Internal Working Paper No. 52 (Limited Circulation). Social Sciences Project, Space Sciences Laboratory, University of California, Berkeley, Calif., September.

PLATT, J. 1969. "What We Must Do," *Science*, Vol. 166 (November 28), pp. 1115–1121.

QUINE, W. V. O. 1953. "Two Dogmas of Empiricism," in *From a Logical Point of View*. Cambridge, Mass.: Harvard University Press.

ROSENBLEUTH, A., BIGELOW, J., and WIENER, N. (1943). "Behavior, Purpose and Teleology," *Philosophy of Science*, 10, No. 1 (January).

SACKMAN, H. 1967. *Computers, Systems Science, and Evolving Society: The Challenge of Man-Machine Digital Systems*. New York: John Wiley & Sons, Inc.

SAMUEL, A. L. 1963. "Some Studies in Machine Learning Using the Game of Checkers," in E. Feigenbaum and J. Feldman, *Computers and Thought*. New York: McGraw-Hill Book Co.

SIMMONS, R. F. 1960. "Anticipated Developments in Machine Literature Processing in the Next Decade," SP–129. Santa Monica, Calif.: System Development Corporation.

SINGER, E. A., Jr. 1924. *Mind as a Behavior*. Columbus: R. G. Adams & Co., p. 282.

————. 1936. *On the Contented Life.* New York: Henry Holt & Co.

————. 1959. *Experience and Reflection,* C. W. Churchman, ed. Philadelphia: University of Pennsylvania Press.

SPINOZA, B. *On the Improvement of the Understanding.*

VICKERS, G. 1970. *Value Systems and Social Process.* Baltimore, Md.: Penguin Books.

INDEX

absolute mind, 70; in Hegel, 178
abstraction: of Democritean imagery, 210
Ackoff, R. L., viii, 47, 121, 279
adaptive systems, 63
adjustment, of readings, 195–196
advisory panels: of scientists, 221
aerospace companies, 181
agnosticism, 240
agreement, 126, 157, 198–199
alienation, in experiments, 159
Anaxagoras, 41, 78
antiteleology, 247–258
antiplanning, 49
a priori, arithmetic, 129–130; design problems, 142; generalizations, 109, 130, 145; geometry, 128; kinematics, 128; language, 142; in Lockean inquiring systems, 127; maximal, 141; in measurement, 194; necessity of, 129; number theory, 128; problem solution, 138; reflection, 129; representation, 140; science, 133; simplicity, 137; validation of, 129
apperception, 75, 94; in Leibniz, 30; in the sciences, 198
Aquinas, St. Thomas, 18
archetype, 244
Aristotle, 210–211, 253, 258
Aristotelian imagery, 210–211
arithmetic, a priori, 129–130
art and uniqueness, 267
artificial, 257
astrology, 244
ateleology, as basic design, 252–255
auditors, and information, 162
authority, and information, 164
automated biology laboratory, 116
automation, of conventional Lockean inquirers, 115

authority, in measurement, 196
authorization, of problem, 143

basic research, 244; as a system, 57
Bayesian approach, 90; probability, 212
beauty, and inquiry, 264
behavioral science, 225
benefit, cost, 163–165, 263; vs. morality, 250
Bennis, W., vii, 279
Berkeley, G., 35
Betz, F., viii
bias, 141
Bigelow, J., 281
Boulding, K., 78, 279
Braybrooke, D., 65, 281
Buchanan, B., 79
Buddha, 204
bureaucracy, and information, 162

calibration, design of, 152
Campbell, J., 203
Carmichael, H., 180
Carneadean imagery, 211
Carneades, 211
causality, 44, 113; in Hume, 130
change: resistance to, 14
checkers, 138
chess, 138
Chomsky, N., 36, 279
circumambulation, 205, 244
client, 42–78, 91, 200
clock, 131, 132; as an a priori, 110
code of conduct, and science, 219
collective unconscious, 245
college: as a system, 57
communication, 5, 60, 61, 123; in Lockean inquiring systems, 123
community: Lockean 97, 101, 154
components, 42–78, 92, 200

283